Allelopathy in Crop Production

Allelopathy in Crop Production

Ajay Mehta

RANDOM PUBLICATIONS
NEW DELHI - 110 002 (INDIA)

Allelopathy in Crop Production

ISBN 978-93-51112-73-0

Published in 2014 in India by

RANDOM PUBLICATIONS

4376-A/4B, Gali Murari Lal, Ansari Road
New Delhi-110 002
Phone: +9111-43580356, 23289044
E-mail: randomexports@gmail.com; sales@randompublications.com; info@randompublications.com

Reprinted 2025

Type Setting by: Friends Media, Delhi-110089
Printed at : Replika Press Pvt. Ltd.

Preface

Allelopathy is the direct influence of a chemical released from one living plant on the development and growth of another. Many researchers have speculated that allelopathy might prove useful in controlling weeds and increasing grain yields. Selection for genotypes with enhanced allelopathic potentials has been carried out in several field crops, and evidence has accumulated that crop cultivars differ significantly in their ability to inhibit the growth of certain weed species. Although traditional breeding methods have not been successful in producing highly allelopathic grain crops with good yields, genetic engineering has the potential for overcoming this impasse. Conceivably, genetic engineers could enhance the production of allelochemicals already present in a crop or impart the ability to produce new compounds. With either strategy, there are potential problems that must be overcome with regard to tissue-specific promoters, autotoxicity, metabolic imbalances, and proper movement of the allelopathic compound to the rhizosphere. Thus, there is a considerable amount of research to do in all the subdisciplines surrounding this problem. Allelopathy is characteristic of certain plants, algae, bacteria, coral, and fungi. Allelopathic interactions are an important factor in determining species distribution and abundance within plant communities, and are also thought to be important in the success of many invasive plants. For specific examples, see Spotted Knapweed (Centaurea maculosa), Garlic Mustard (Alliaria petiolata), Casuarina/Allocasuarina spp, and Nutsedge. The process by which a plant acquires more of the available resources from the environment without any chemical action on the surrounding plants is called resource competition. This process is not negative allelopathy, although both processes can act together to enhance the survival rate of the plant species.

Strategies for utilizing allelopathy as an aid in crop production include both avoidance and application protocols. There are immediate opportunities for management of weed and crop residues, tillage practices, and crop sequences to minimize crop losses from allelopathy and also to use allelopathic crops for weed control. Strip cropping that included sorghum showed that in the subsequent year weed density and biomass were significantly lower in the previous-year sorghum than in soybean strips. Possibilities exist for modification of crop plant metabolism to alter production of allelochemicals. Allelochemical-environmental interactions must be considered in efforts to benefit from allelopathy. Under greenhouse conditions, joint application of low levels of atrazine, trifluralin, alachlor, or cinmethylin with a phenolic allelochemical showed that these two categories of inhibitors acted in concert to reduce plant growth. Allelochemicals may also be adapted as yield stimulants or environmentally sound herbicides, such as cinmethylin and methoxyphenone. Isolation of bialophos, tentoxin, and others shows that bacteria and fungi are good sources of biologically active compounds.

The book provides the students current information on the different areas of this subject. The publication is useful not only to the students but also to the research scholars and academic professionals.

I thank all members of my team who have helped in the preparation of the book. My special thanks go to "Random Publications" who have published the book.

—Ajay Mehta

Contents

1

Introduction

Allelopathy

Figure: Casuarina equisetifolia *litter completely suppresses germination of understory plants as shown here despite the relative openness of the canopy and ample rainfall (>120 cm/yr) at the location*

Allelopathy is a biological phe nomenon by which an organism produces one or more biochemicals that influence the growth, survival, and reproduction of other organisms. These biochemicals are known as allelochemicals and can have beneficial (positive allelopathy) or detrimental (negative allelopathy) effects on the target organisms.

Allelochemicals are a subset of secondary metabolites, which are not required for metabolism (i.e. growth, development and reproduction) of the allelopathic organism. Allelochemicals with negative allelopathic effects are an important part of plant defence against herbivory.

Allelopathy is characteristic of certain plants, algae, bacteria, coral, and fungi. Allelopathic interactions are an important factor in determining species distribution and abundance within plant communities, and are also thought to be important in the success of many invasive plants.

The process by which a plant acquires more of the available resources (such as nutrients, water or light) from the environment without any chemical action on the surrounding plants is called resource competition. This process is not negative allelopathy, although both processes can act together to enhance the survival rate of the plant species.

History

The term allelopathy from the Greek-derived compounds *allelo-* and *-pathy* (meaning "mutual harm" or "suffering"), was first used in 1937 by the Austrian professor Hans Molisch in the book *Der Einfluss einer Pflanze auf die andere - Allelopathie* (The Effect of Plants on Each Other) published in German. He used the term to describe biochemical interactions that inhibit the growth of neighbouring plants, by another plant. In 1971, Whittaker and Feeny published a study in the journal *Science*, which defined allelochemicals as all chemical interactions among organisms. In 1984, Elroy Leon Rice in his monograph on allelopathy enlarged the definition to include all direct positive or negative effects of a plant on another plant or on by the liberation of biochemicals into the natural environment. Over the next ten years, the term was used by other researchers to describe broader chemical interactions between organisms, and by 1996 the International Allelopathy Society (IAS) defined allelopathy as "Any process involving secondary metabolites produced by plants, algae, bacteria and fungi that influences the growth and development of agriculture and biological systems." In more recent times, plant researchers have begun to switch back to the original definition of substances that are produced by one plant that inhibit another plant. Confusing the issue more, zoologists have borrowed the term to describe chemical interactions between invertebrates like corals and sponges.

Long before the term allelopathy was used, people observed the negative effects that one plant could have on another. Theophrastus, who lived around 300 BC noticed the inhibitory effects of pigweed on alfalfa. In China around the first century AD, the author of *Shennong Ben Cao Jing* described 267 plants that had pesticidal abilities, including those with allelopathic effects. In 1832, the Swiss botanist De Candolle suggested that crop plant exudates were responsible for an agriculture problem called soil sickness.

Allelopathy is not universally accepted among ecologists and many have argued that its effects cannot be distinguished from the competition which results when two (or more) organisms attempt to use the same limited resource, to the detriment of one or both. Allelopathy is a direct negative effect on one organism resulting from the input of substances into the environment by another. In the 1970s, great effort went into distinguishing competitive and allelopathic effects by some researchers, while in the 1990s others argued that the effects were often interdependent and could not readily be distinguished.

Examples of Allelopathy

The possible application of allelopathy in agriculture is the subject of much research. Current research is focused on the effects of weeds on crops, crops on weeds, and crops on crops. This research furthers the possibility of using allelochemicals as growth regulators and natural herbicides, to promote sustainable agriculture. A number of such allelochemicals are commercially available or in the process of large-scale manufacture. For example, Leptospermone is a purported thermochemical in lemon bottlebrush (*Callistemon citrinus*). Although it was found to be too weak as a commercial herbicide, a chemical analog of it, mesotrione (tradename Callisto), was found to be effective. It is sold to control broadleaf weeds in corn but also seems to be an effective control for crabgrass in lawns. Sheeja (1993) reported the allelopathic interaction of the weeds *Chromolaena odorata* (*Eupatorium odoratum*) and *Lantana camara* on selected major crops.

Many crop cultivars show strong allelopathic properties, of which rice (*Oryza sativa*) has been most studied. Rice allelopathy depends on variety and origin: Japonica rice is more allelopathic than Indica and Japonica-Indica hybrid. More recently, critical review on rice allelopathy and the possibility for weed management reported that allelopathic characteristics in rice are quantitatively inherited and several allelopathy-involved traits have been identified.

Many invasive plant species interfere with native plants through allelopathy. A famous case of purported allelopathy is in desert shrubs.

One of the most widely known early examples was *Salvia leucophylla*, because it was on the cover of the journal *Science* in 1964. Bare zones around the shrubs were hypothesized to be caused by volatile terpenes emitted by the shrubs. However, like many allelopathy studies, it was based on artificial lab experiments and unwarranted extrapolations to natural ecosystems. In 1970, *Science* published a study where caging the shrubs to exclude rodents and birds allowed grass to grow in the bare zones. A detailed history of this story can be found in Halsey 2004.

Allelopathy has been shown to play a crucial role in forests, influencing the composition of the vegetation growth, and also provides an explanation for the patterns of forest regeneration. The black walnut *(Juglans nigra)* produces the allelochemical juglone, which affects some species greatly while others not at all. The leaf litter and root exudates of some *Eucalyptus* species are allelopathic for certain soil microbes and plant species. The tree of heaven, *Ailanthus altissima*, produces allelochemicals in its roots that inhibit the growth of many plants. The pace of evaluating allelochemicals released by higher plants in nature has greatly accelerated, with promising results in field screening.

Garlic mustard is an invasive plant species in North American temperate forests. Its success may be partly due to its excretion of an unidentified allelochemical that interferes with mutualisms between native tree roots and their mycorrhizal fungi.

A study of *Kochia scoparia* in northern Montana by two high school students showed that when *Kochia* precedes spring wheat (*Triticum aestivum*), it reduces the spring wheat's growth. Effects included delayed emergence, decreased rate of growth, decreased final height and decreased average vegetative dry weight of spring wheat plants. A larger study later showed that *Kochia* seems to exhibit allelopathy on various crops in northern Montana.

Corn gluten meal (CGM) is a natural preemergence weed control used in turfgrass, which reduces germination of many broadleaf and grass weeds.

The Importance of Allelopathy in Breeding New Cultivars

Allelopathy is defined as the direct influence from a chemical released from one plant on the development and growth of another. It is known that allelopathic substances are induced by environmental stresses. Allelopathic compounds may be released into the environment from plants by means of root exudation, leaching, volatilization and decomposition of plant residues in the soil. Allelopathic substances, if present in crop varieties, may reduce the need for weed management, particularly herbicide use. Allelopathy alone may not be a perfect weed management technology but it may be a supplementary tool for weed control. It is extremely difficult to demonstrate allelopathy in nature because of the complexity of plant interference which includes positive, negative and neutral effects on each other (Christensen, 1993). Interference is a combination of the processes of competition for resources and production of allelopathic compounds which suppress competitors (Duke *et al.* 2001). Thus, allelopathy differs from resource competition. Alleopathy research has been conducted for several decades, but very limited knowledge is still available. An improvement in crop cultivars is the only area that has not been exploited to any great extent as a weed management strategy (Khush, 1996). The possibility of incorporating allelopathic traits into improved rice cultivars, which would reduce the need for applying herbicides to the crop, is worth exploring (Khush, 1996). Of course, thus far, no commercial cultivars carrying allelopathic properties have been developed (Duke *et al.* 2001).

Recently, the allelopathic potential of rice has received a great deal of attention since Dilday *et al.* (1991) identified rice cultivars exhibiting allelopathic potential against ducksalad [*Heteranthera limosa* (Sw.) Willd]. In addition, allelopathic potential has been also reported from numerous crops like barley (Lovett and Hoult, 1995); cucumber (Putnam and Duke, 1974); oats (Fay and Duke 1977); rice (Dilday *et al.* 1998); sorghum (Nimbal *et al.* 1996); sunflower (Leather, 1983); tobacco (Patrick *et al.* 1963); and wheat (Wu *et al.* 1999).

Although a breeding approach alone cannot overcome weed problems, an increase in the allelopathic potential of rice varieties will likely have a great impact on both low- and high-input cropping systems. Moreover, allelopathy-based technology is also more easily transferable to farmers in low-input management systems than those in high-input management systems, which entail a heavy use herbicide. There are three excellent reviews on this subject: one is on strategies to enhance weed control by engineered allelochemical production (Duke *et al.* 2001); the other two reviews are by Olofsdotter (2001a, 2001b) on rice allelopathy.

Here will briefly discuss the importance of breeding allelopathic crops, focusing on allelopathic traits, methods of measuring genetic difference, the allelochemicals identified, and an introductory method on how to breed allelopathic crops. Rice was used as the model plant, but it would seem that the principles described are applicable to any other crop.

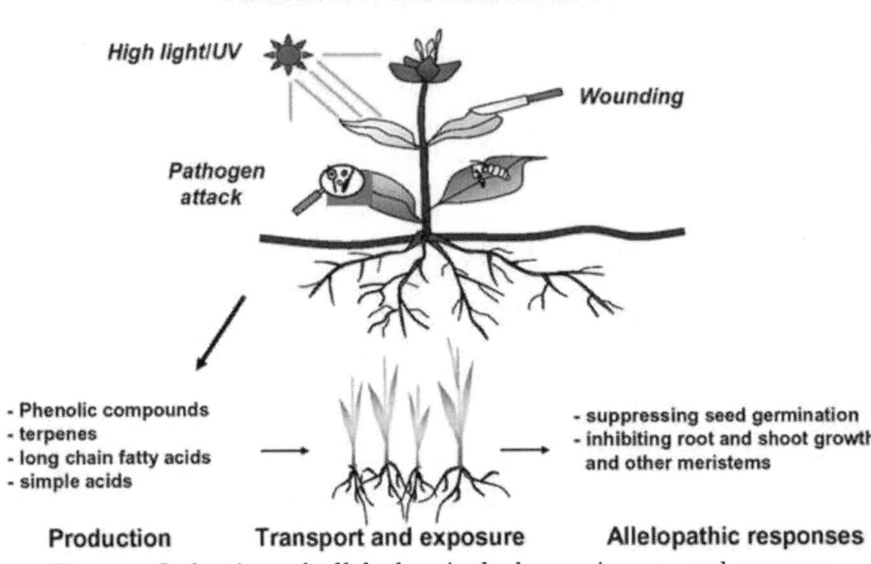

Figure: *Induction of allelochemicals by environmental stresses.*

Importance of Allelopathy in Breeding New Crops

Recent research on rice allelopathy has advanced the possibility of breeding allelopathic rice crops for the following reasons:

- A variation in allelopathy among rice cultivars exists.
- Allelopathy plays a role under field conditions.
- Allelopathic rice is effective on both mono- and dicot weeds.
- Identification of rice allelochemicals has progressed remarkably.
- Quantitative trait loci (QTL) correlated with allelopathy have been determined (Olofsdotter, 2000b).

A number of rice cultivars or accessions having allelopathic potential have been determined in different places (Dilday *et al.* 1991; Fujii, 1992: Kim and Shin, 1998, Olofsdotter *et al.* 1999). However, no commercial allelopathic rice cultivar has been developed yet. In this regard, the problem now is how to utilize such allelopathic potential in breeding programmes. Two very important effects can be expected from the breeding of allelopathic crops.

Ecological Importance

In all rice ecosystems, herbicides have become one of the most important components in weed control. There are two reasons to explain the increased use of herbicides, the first being the widespread adoption of high-yielding varieties which created economic incentives for farmers to reduce weed infestation; and the second is the availability of cheap herbicides, indicating that the cost of weed control by herbicides in wet-seeded rice is less than one-fifth of the cost of a single hand-weeding in Illoilo, Philippines (Moody, 1991). Similar situations exist in West Java, Indonesia and the Mekong Delta, Vietnam (Pandy and Pingali, 1996).

Because of the availability of cheap herbicides, it is expected that herbicide usage will continue to increase, both in developed countries, and even in developing countries, where herbicides are currently used sparingly and farm wages are relatively low. However, this does not indicate a lack of importance for hand-weeding. Manual weeding is still the dominant weed control method in many parts of Asia, since management options for weed control are limited under diverse agro-ecological conditions (Kim, 2000).

All the recommended herbicides commonly used are known to be very safe not only for humans and cattle, but also for the environment when they are used properly. Sulfonylurea type herbicides have been intensively used in far-east Asia since 1990 because of their high efficacy against a broad spectrum of paddy weeds even at extremely low doses. However, intensive and repeated application of this type of herbicide has resulted in several negative effects, as follows:

- evolving resistant weeds (Valverde *et al.* 2000);
- residual effects on the following crops;
- disappearance of some susceptible weeds such as *Brasenia schreberi* and *Sagittaria aginashi,* which affects weed biodiversity (Itoh, 2000).

Another important element is that rice farming in Asia is closely linked with fish farming and irrigation water used for other purposes. To eliminate molinate and thiobencarb pollution of drinking water, over US$600 000 was expended in cost-share assistance alone to implement water recovery systems in rice production in California (Hill *et al.* 1977). This was a shock for rice farmers as well as consumers, strongly suggesting an urgent need to monitor herbicide movement and pollution in paddy fields, rivers, all aquatic systems, and even in non-target organisms.

All these factors may well provide sufficient reason to attract public concern and anxiety regarding the negative effects of herbicides that might originate from intensive herbicide application in the environment. In this regard, an alternative to such a heavy dependence on herbicide is needed. Such an alternative might be found in the use of allelopathy, which can reduce herbicide use in rice cropping systems.

Economical Importance

The use of allelopathic crops can definitely reduce the cost of weed control. This technology, (allelopathy), which is seed-based technology is more easily transferable to the low-input management systems prevailing throughout most Asian rice-farming systems.

In order to implement this technology, the anticipated problems such as autotoxicity, residual effects of allelopathic cultivars and tolerance of weed population upon repeated cultivation of allelopathic cultivars in the same fields should be thoroughly studied before allelopathic cultivars are released to farmers. The establishment of an appropriate weed-management system could surely help to overcome such negative effects.

Allelopathic Traits

What are Allelopathic Traits?

Morphological characteristics such as early seedling emergence, seedling vigour, fast growth rates that produce a dense canopy, greater plant height, greater root volume and longer growth duration are known to increase the ability of rice cultivars to compete with weeds (Minotti and Sweet, 1981, Bekowitz, 1988). Plant height is often described as one of the most important factors in the total competitive ability of a crop and accounts for a similar percentage of total competitive ability (Gaudet and Keddy, 1988; Garrity *et al.* 1992). However, it is not certain that all of these characteristics are also related to allelopathy.

How can we measure such allelopathic effects? If there is no specific characteristic, the most visible plant characteristics such as plant height, root length and dry weight of testing plants can be used as measurable parameters to evaluate allelopathic potential. To breed new allelopathic crops, the first requirement is to identify allelopathic accessions or cultivars. To identify allelopathic cultivars, an effective screening method that can test large amounts of accessions in a limited-space in an easy, cheap, reproducible and fast manner must be established.

Screening Methods

To screen rice allelopathic potential, several methods such as the stairstep method (Bonner, 1950; Liu and Lovett, 1993), hydroponic culture test (Einhelig, 1985), relay-seeding technique (Navarez and Olofsdotter, 1996), agar medium test (Fujii, 1992; Wu *et al.* 1999), cluster analysis using HPLC (Mattice *et al.* 2001), water extract method (Kim *et al.* 1999; Ebana *et al.* 2001) and 24-well plate bioassay (Rimando *et al.* 1998) have been suggested. Each of the methods mentioned above can be used as a valuable tool to evaluate allelopathic potential. However, it has also been observed that the screened accessions or cultivars in one method may not be always active in another method employed. In terms of screening, the direct field test might be the best method to evaluate allelopathic potential, but it has many difficulties when dealing with a large number of germplasms. Thus, what is actually needed is a reliable and universal bioassay method that generates a similar trend under different conditions. For this purpose, the continued development of reliable screening systems is needed.

Genetic Variability

Different Allelopathic Potential

Dilday *et al.* (1998) identified 412 accessions having allelopathic potential against ducksalad [*Heteranthera limosa* (Sw.) Willd] among 12 000 accessions that originated from 31 different countries. These accessions were genetically very diverse, indicating that allelopathic potential is widely present in rice germplasm. A number of studies have been conducted to evaluate the allelopathic potential from rice germplasm. A large variation in allelopathic potential among rice cultivars has been reported to exist in germplasms. Further, it is assumed that allelopathic potential might be polygenically controlled because it shows a continuous variation in the germplasm. Moreover, allelopathic potential is often attributed to several inhibitors that are assumed to act in an additive or synergistic way rather than in an isolated way (Courtois and Olofsdotter, 1998). Tropical *japonica* rice varieties have been shown to have a greater allelopathic potential against weeds than other rice, especially *Echinochloa* spp. (Jensen *et al.* 2001).

Quantitative Trait Loci Analysis

Very limited knowledge exists on allelopathy genetics. Moreover, no intentional breeding effort has been made to genetically improve the allelopathic potential of crops, mainly because of our poor knowledge of this phenomenon. In this regard, recent work conducted by Jensen *et al.* (2001) seems very valuable. Quantitative trait loci (QTL) mapping using 142 DNA markers were located in 142 recombinant inbred lines derived from a cross between cultivar IAC 165 (*japonica* upland variety), which has strong allelopathic potential and cultivar CO 39 (*indica* irrigated variety), which has weak allelopathic potential. Three main loci, each accounting for about 10 percent of the upregulation of allelochemical production were localized to rice chromosomes 2 and 3. The two QTL traits on chromosome 3 were closely linked, so they could easily be manipulated. Although QTLs for allelopathic effect against barnyardgrass were identified, it is not presently known what kinds of gene are responsible for the allelopathic effect or the production of allelochemcals.

Rice Allelochemicals

Allelochemical Synthesis Affected by Environmental Stresses

Until recently, many studies verified the mechanisms of a self-defence system, including allelopathy in plants, particularly phenylpropanoid, and isoterpenoid metabolism. Plants respond to environmental stress through a variety of biochemical reactions, which may provide protection against casual agents. The increase of allelopathic phenolic and terpenoid compounds under environmental stresses has been well documented. For example, enhanced UV-B light induces the accumulation of phenylpropanoids and flavonoids in different plant species, such as bean, parsley, potato, tomato, maize, rye, barley and rice (Hahlbrock and Scheel 1989; Ballare *et al.* 1995; Tevini *et al.* 1991; Liu *et al.* 1995; Kim *et al.* 2000a).

Pathways for Allelochemical Synthesis

All phenylpropanoids are derived from cinnamic acid, which is formed from phenylalanine by the catalytic action of phenylalanine ammonia-lyase (PAL), the branch point enzyme between primary (shikimate pathway) and secondary (phenylpropanoid) metabolism. It is known that many phenolic compounds not only have a physiologically functional ability, but also plant allelopathic potential.

Isoprenoid compounds are produced from C_5 isoprenoid units and the classification of different families of isoprenoids is based on the number of C_5 isoprenoid units present in the skeleton of compounds (Gershenzon and Croteau, 1993). In particular, diterpenoids are known to play an important role in the self-defence mechanism of rice as well as alelopathic potential. Terpenoids play diverse functional roles in plants as hormones (gibberellins, abscisic acid), photosynthetic pigments (phytol, carotenoids), electron carriers (ubiquinone, plastophates), mediators of polysaccharide assembly (polyprenyl phosphates), and structural components of membranes (phytosterols). In addition to the universal physiological, metabolic and structural functions, many specific terpenoid compounds (commonly in the C_{10}, C_{15}, and C_{20} families) serve in communication and defence. It has been determined that induced specific terpenoids have been correlated with plant-plant, plant-insect, and plant-pathogen interaction

Putative Allelochemicals

There are a number of studies indicating that common putative allelopathic substances found in rice are phenolic compounds (Rice, 1987; Chou *et al.* 1991; Inderjit, 1996; Mattice *et al.* 1998; Blum, 1998). Kim *et al.* (2000b) identified several compounds by GC/MS analysis from Kouketsumochi: potential allelopathic rice, such as sterols, benzaldehydes, benzene derivatives, long-chain fatty acid esters, aldehydes, ketones and amines from fractions with biological activity. Several compounds from Taichung Native 1, allelopathic rice, have been identified by the bioassay-guided isolation method (Rimando *et al.* 2001). They are azelaic acid, *p*-coumaric acid, 1*H*-indole-carboxaldehyde, 1*H*-indole-3-carboxylic acid, 1*H*-indole-5-carboxylic acid and 1,2-benzenedicarboxylic acid bis (2-ethylhexyl) ester. Among the allelopathic substances identified, *p*-coumaric acid, a known allelochemical, inhibited the germination of lettuce (*Lactuca sativa* L.) seedlings at 1m*M*, but was active against barnyardgrass only at concentrations higher than 3 m*M* (Rimando *et al.* 2001). A similar trend in results was reported by Kim *et al.* (2001b) in that *p*-coumaric acid was present in potential allelopathic rice such as Kouketsumochi, Tang Gan and Taichung Native 1. In addition, the content of this acid was markedly increased by UV irradiation.

These results strongly suggest that *p*-coumaric acid may be a compound regulated by environmental stresses and play a role in rice allelopathy. On the contrary, there were different reports that concentration of a single phenolic acid and combinations of all phenolic acids determined in rice ecosystems do not reach phytotoxic levels (Tanaka *et al* 1990; Olofsdotter *et al.* 2001). Thus far, it is assumed that no single allelochemical responsible for rice allelopathy has been determined. On the contrary, more than one chemical is likely to be responsible for allelopathy in rice. Thus, the identification of allelochemicals in rice is of utmost importance for use as markers in gene identification and regulation.

Strategies For Breeding Allelopathic Crops

The three approaches to create more allelopathic crops are the traditional breeding method; the incorporation of allelopathic properties to hybrid rice; and genetic engineering. From the breeding perspective, the distinction between allelopathy and competition is somewhat difficult in a farmer's field because interference is the phenomenon that really counts (Courtois and Olofsdotter, 1998). In this regard, the best way might be to breed allelopathic cultivars having high competitiveness. If breeding of allelopathic cultivars is achieved, it can be truly beneficial to farmers in rice-producing countries.

Traditional Breeding

If a high number of QTLs with little effect are involved, a traditional breeding method can be a reasonable alternative (Courtois and Olofsdotter, 1998). The principle of traditional breeding for the genetic studies is simple. Two parents with contrasting behaviour are crossed and recombinant inbred lines (RILs) are derived through the single-seed descent (SSD) method.

This procedure consists of advancing the F_2 without selection for two or three generations (sometime four to five generations) in such a way that each F_4 or F_5 seed traces back to a different F_2 plant. Only one seed is retained from each plant in each generation. This will repeat for two or more generations so that afterwards the plants having desirable characteristics can be selected. Once a reasonable degree of fixation is obtained, the allelopathic potential of the RILs can be evaluated. The seed can then be increased and tested under field conditions for use as allelopathic cultivars.

A cross between Donginbyeo (non-allelopathic cultivar, but a high yielding variety with good quality) and Kouketsumochi (an allelopathic cultivar, more similar to a wild type) was made and advanced by the SSD breeding method. We identified the F_5 generation of this cross-exhibited allelopathic potential when determined by the water-extract method. The allelopathic potential of the F_6 generation is now being tested under field conditions. The SSD method is simpler than the pedigree method. The SSD method has several advantages, including a quick increase of the additive variance among families; the need for only small breeding space and saving of work; and suitability for low heritability traits where visual selection is not effective (Moreno-González and Cubero, 1993).

Another method is to cross two parents with contrasting behaviour through the backcrossing method to produce near-isogenic lines (NILs) carrying different genes. The allelopathic potential of the NILs can be determined when a reasonable degree of fixation is obtained.

Despite the influence of biotechnology and genetic engineering in changing future production methods for breeding new crop varieties, traditional plant breeding research and breeding methodology will play a significant role in future breeding programmes. Further refinement of these methods and better knowledge of classical knowledge are a prerequisite for the rational use of new tools such as molecular markers.

Hybrid Rice with Allelopathy

The three-line hybrid rice that is popularly cultivated in China is a good competitor because of its rapid and profuse vegetative growth in comparison with an inbred line. It will be worthwhile if allelopathic traits can be introduced into an elite restorer line in developing three-line hybrid rice. Lin *et al.* (2000) attempted simultaneous backcrossing and selfing breeding method to develop hybrid rice with allelopathic activity and its counterpart, isogenic hybrid rice with a non-allelopathic effect on weeds. Three lines of rice, Kouketsumochi, Rexmont and IR24 were used as the donors of allelopathy, non-allelopathy and restoring genes, respectively. The back-crossing method was used to develop restorer lines with allelopathic activity and its isogenic line with non-allelopathic genes. The selected restorer lines were crossed with cytoplasm-sterile lines (A line), presently used to test the outcross rate and also illustrates a scheme on how to develop hybrid rice having allelopathic potential.

Allelopathic potential and heterotic performance were analysed in a laboratory and greenhouse using the water-extract method and density-dependent method, respectively (Richardson *et al.* 1988; Lin *et al.* 2000). The results indicated that the heterotic effect on allelopathy in rice was positively significant, showing higher heterosis over the mid-parent. In addition, this specific hybrid rice had an additive suppressive effect on the target weed of barnyardgrass, exhibiting a large deviation from the resource competition curve (Lin *et al.* 2000). This approach seems to be a promising method that can incorporate allelopathic traits to hybrid rice having high yield capacities.

Molecular Approach

Two methods for creating more allelopathc crops have been suggested: (1) the regulation of gene expression related to alleochemical biosynthesis; and (2) the insertion of genes to produce allelochemicals that are not found in the crop (Duke *et al.* 2001). Much of the presently available information on engineering allelochemicals comes to us through efforts to overproduce valuable secondary metabolites in plants (Canel, 1999). Relating to the molecular approach for weed control, a good reference has recently been published by Gressel (2002).

Most secondary metabolites being used as allelochemicals are very complex and a multi-gene system might have to be developed and transformed into the specific crop to produce allelochemicals (Gressel, 2002). This seems to be still hypothetical, but new areas opening up. Various plant species suppress other species by production of allelochemicals, which are not toxic to the originating plant but toxic to other vegetation. It is now known that molecular approaches in breeding allelopathic cultivars are potentially much more complicated than developing a herbicide-resistant crop or producing a crop with resistance to insects or pathogens.

Regulation of gene Expression related to Allelochemicals

In order to regulate gene expression, the first requirement is to identify the target allelochemical(s), to determine enzymes and the genes encoding them, and thus a specific promoter can be inserted to crop plants to enhance allelochemical production.

In our recent work, the highest content of *p*-coumaric acid was found in among intermediates of the phenylpropanoid pathway of several allelopathic rice cultivars used. Cinnamic acid 4-hydroxylase (CA4H) is the enzyme catalyzing cinnamic acid to *p*-coumaric acid, which is a key reaction in the biosynthesis of a large number of phenolic compounds in higher plants. The activity of CA4H was measured to elucidate how the activity is influenced by UV irradiation in rice leaves. CA4H activity in Kouketsumochi was induced by UV and its peak activity was observed at 24 h after UV irradiation for 20 min. AUS 196 showed no response, however, indicating a differential response to UV or other environmental stresses among rice cultivars. This indicates further that some metabolites having allelopathic potential might be newly synthesized or highly elevated in rice plants by UV irradiation (Kim *et al.* 2000a). Owing to the increased content of *p*-coumaric acid in rice plant irradiated by UV, it was assumed that the CA4H gene plays a role in the elevation of the allelopathic function in rice plants.

Further study was conducted to investigate a specific promoter that confers responsiveness on environmental stresses and plant-

plant interaction. If a promoter is specifically responsive to an elicitor, it can be used to regulate genes to code allelopathic potential. We examined the expression of GUS activity of tobacco plants transformed with the constructs of various CASC (*Capsicum annuum* sesqiterpene cyclase) promoters fused into GUS gene in order to determine the effective motifs response to UV light. Back *et al.* (1998) reported that UV plays a role in the expression of defence-related genes and biosynthesis of phytoalexins in pepper.

To determine whether the CASC promoters are inducible by UV irradiation or not, we examined GUS activity of the transgenic plants. Transgenic plants were UV irradiated for 15 min using a Hitachi germicidal lamp (15W) at a distance of 20 cm. The irradiated leaves were placed on a MS medium for 24 h and were used for the activity measurement. The levels of GUS activity for transgenic plants with pBI121-KF1 and pBI121-KF6 were significantly elevated by UV-irradiation and had a two-to-threefold increase approximately over the untreated-transgenic plants. In contrast, GUS expression in the transgenic plants with pBI121-CaMV 35S was not changed by UV, and in the other constructs had only a very small increase (Shin *et al.* 2000).

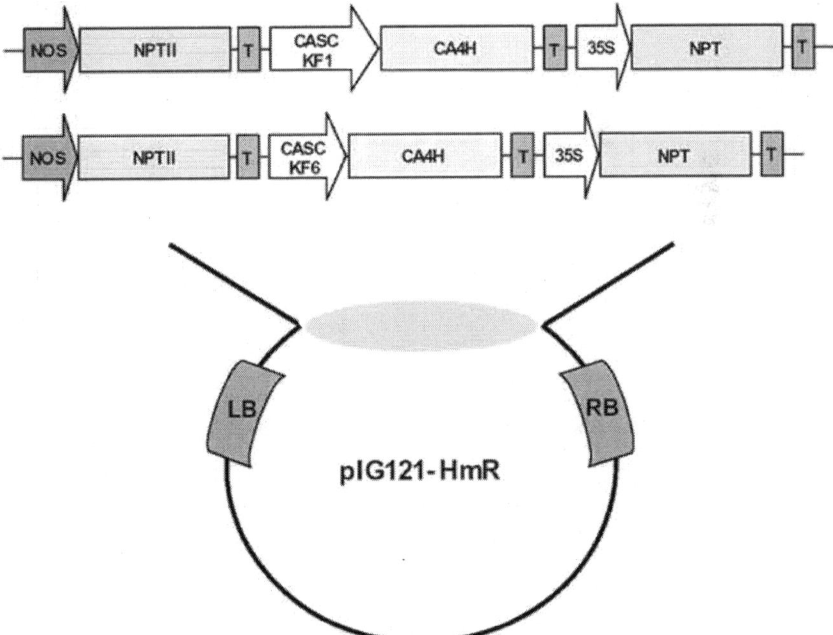

Figure: *The gene cassette with specific promoters responsive to UV irradiation in pIG121-HmR.*

This result suggests that the CASC promoters of KF-1 and KF-6 may contain cis-acting elements capable of conferring quantitative expression patterns that are exclusively associated with UV irradiation. The regulation of genes associated with allelopathy can be achieved by developing a specific promoter responsive to plant-weed competition or environmental stresses. The CASC promoters of KF-1 and KF-6 obtained may be specific to UV. Thus, this promoter can be used for the overexpression of specific promoters constructed to allelochemical-producing genes (Shin *et al.*). To regulate the CA4H gene in the phenylpropanoid pathway, specific promoters, the CASC-KF1 and KF6 were fused to CA4H gene. The gene constructs were introduced into the binary plant expression vector pIG121-HMR with reverse primer harbouring *Bam*HI site and forward primer harbouring *Hin*dIII site. We are at present investigating the regulatory effects on gene coding for allelochemical-producing enzymes.

Insertion of Genes to Produce Allelochemicals

This approach employs altering existing biochemical pathways by the insertion of transgenes to produce new allelochemicals. It seems to be a difficult approach to employ, but it remains one of the promising molecular approaches to be advanced in the near future. In this regard, there was a good review conducted by Duke *et al.* (2001) and a reference book on molecular biology of weed control very recently published by Gressel, (2002).

An improvement in allelopathic potential in crop varieties will have a great impact on both low-input and high-input management systems. Allelopathy alone is not likely to replace totally other weed control practices because its effectiveness is influenced by many factors. However, marginally reduced use of herbicide over time will be a significant economical benefit to farmers and will also reduce the ecological impact on the environment. At present, no commercial cultivars carrying allelopathic properties are available, but there is the possibility of breeding new allelopathic crops by regulating their capacity to produce allelopathic substances. In association with the development of breeding new allelopathic crop cultivars, much progress has been made in screening methods, identification of specific allelochemicals and related genes that can be applicable in breeding programmes. If allelochemicals or genes responsible for allelopathic effects are identified, allelopathic traits can be easily incorporated into improved cultivars through breeding techniques available at the present time. Of course, in order for this technology to be practically used in rice production systems, potential problems such as autotoxicity, metabolic imbalance, residual effect and the development of a tolerant population on repeated cultures of such cultivars should be thoroughly solved before allelopathic cultivars are released. Nevertheless, it is assumed that the breeding of new allelopathic crop cultivars will represent an attractive alternative in meeting the demands of consumers and environmentalists. A multidisciplinary team approach among a wide range of scientists, including weed scientists, ecologists, natural-product chemists, plant breeders and molecular biologists, could help to achieve the goal of breeding allelopathic crop cultivars. Further studies on the identification of specific allelochemicals and genes responsible for allelopathy will shed more light on breeding allelopathic crop cultivars.

2

Agricultural use of Sewage Sludge

Characteristics of Sewage Sludge

Most wastewater treatment processes produce a sludge which has to be disposed of. Conventional secondary sewage treatment plants typically generate a primary sludge in the primary sedimentation stage of treatment and a secondary, biological, sludge in final sedimentation after the biological process. The characteristics of the secondary sludge vary with the type of biological process and, often, it is mixed with primary sludge before treatment and disposal. Approximately one half of the costs of operating secondary sewage treatment plants in Europe can be associated with sludge treatment and disposal. Land application of raw or treated sewage sludge can reduce significantly the sludge disposal cost component of sewage treatment as well as providing a large part of the nitrogen and phosphorus requirements of many crops.

Very rarely do urban sewerage systems transport only domestic sewage to treatment plants; industrial effluents and storm-water runoff from roads and other paved areas are frequently discharged into sewers. Thus sewage sludge will contain, in addition to organic waste material, traces of many pollutants used in our modern society. Some of these substances can be phytotoxic and some toxic to humans and/or animals so it is necessary to control the concentrations in the soil of potentially toxic elements and their rate of application to the soil. The risk to health of chemicals in sewage sludge applied to land has been reviewed by Dean and Suess (1985).

Sewage sludge also contains pathogenic bacteria, viruses and protozoa along with other parasitic helminths which can give rise to potential hazards to the health of humans, animals and plants. A WHO (1981) Report on the risk to health of microbes in sewage sludge applied to land identified salmonellae and *Taenia* as giving rise to greatest concern. The numbers of pathogenic and parasitic organisms in sludge can be significantly reduced before application to the land by appropriate sludge treatment and the potential health risk is further reduced by the effects of climate, soil-microorganisms and time after the sludge is applied to the soil. Nevertheless, in the case of certain crops, limitations on planting, grazing and harvesting are necessary.

Apart from those components of concern, sewage sludge also contains useful concentrations of nitrogen, phosphorus and organic matter. The availability of the phosphorus content in the year of application is about 50% and is independent of any prior sludge treatment. Nitrogen availability is more dependent on sludge treatment, untreated liquid sludge and dewatered treated sludge releasing nitrogen slowly with the benefits to crops being realised over a relatively long period. Liquid anaerobically-digested sludge has high ammonia-nitrogen content which is readily available to plants and can be of particular benefit to grassland. The organic matter in sludge can improve the water retaining capacity and structure of some soils, especially when applied in the form of dewatered sludge cake.

The application of sewage sludge to land in member countries of the European Economic Commission (EEC) is governed by Council Directive No. 86/278/EEC (Council of the European Communities 1986). This Directive prohibits the sludge from sewage treatment plants from being used in agriculture unless specified requirements are fulfilled, including the testing of the sludge and the soil. Parameters subject to the provisions of the Directive include the following:

- Dry matter (%)
- Organic matter (% dry solids)
- Copper (mg/kg dry solids)
- Nickel (mg/kg dry solids)
- pH
- Nitrogen, total and ammoniacal (% dry solids)
- Phosphorus, total (% dry solids)
- Zinc (mg/kg dry solids)
- Cadmium (mg/kg dry solids)
- Lead (mg/kg dry solids)

- Mercury (mg/kg dry solids)
- Chromium (mg/kg dry solids)

To these parameters the UK Department of the Environment (1989) has added molybdenum, selenium, arsenic and fluoride in the recent 'Code of Practice for Agricultural Use of Sewage Sludge'. Sludge must be analyzed for the Directive parameters at least once every 6 months and every time significant changes occur in the quality of the sewage treated. The frequency of analysis for the additional four parameters may be reduced to not less than once in five years provided that their concentrations in the sludge are consistently no greater than the following reference concentrations: Mb - 3mg/kg dry solids, Se - 2mg/kg dry solids, As -2mg/kg dry solids and Fl - 200mg/kg dry solids.

Sludge Treatment

Except when it is to be injected or otherwise worked into the soil, sewage sludge should be subjected to biological, chemical or thermal treatment, long-term storage or other appropriate process designed to reduce its fermentability and health hazards resulting from its use before being applied in agriculture. Table below lists sludge treatment and handling processes which have been used in the UK to achieve these objectives. The second edition of a 'Manual of Good Practice on Soil Injection of Sewage Sludge' has been produced by the Water Research Centre (1989) in the UK and describes suitable equipment and techniques for what is now the only method permissible within the EEC for applying untreated sludges to grassland.

Table: *Examples of Effective Sludge Treatment Processes*

Process	***Descriptions***
Sludge Pasteurization	Minimum of 30 minutes at 70°C or minimum of 4 hours at 55° C (or appropriate intermediate conditions), followed in all cases by primary mesophilic anaerobic digestion
Mesophilic Anaerobic Digestion	Mean retention period of at least 12 days primary digestion in temperature range 35°C +/- 3°C or of at least 20 days primary digestion in temperature range 25°C + /- 3°C followed in each case by a secondary stage which provides a mean retention period of at least 14 days

Contd...

Process	***Descriptions***
Thermophilic Aerobic Digestion	Mean retention period of at least 7 days digestion. All sludge to be subject to a minimum of 55°C for a period of at least 4 hours
Composting (Windrows or Aerated Piles)	The compost must be maintained at 40°C for at least 5 days and for 4 hours during this period at a minimum of 55°C within the body of the pile followed by a period of maturation adequate to ensure that the compost reaction is substantially complete
Lime Stabilization of Liquid Sludge	Addition of lime to raise pH to greater than 12.0 and sufficient to ensure that the pH is not less than 12 for a minimum period of 2 hours. The sludge can then be used directly
Liquid Storage	Storage of untreated liquid sludge for a minimum period of 3 months
Dewatering and Storage	Conditioning of untreated sludge with lime or other coagulants followed by dewatering and storage of the cake for a minimum period of 3 months if sludge has been subject to primary mesophilic anaerobic digestion, storage to be for a minimum period of 14 days

Source: *Department of the Environment (1989)*

Sludge Application

The concentrations of potentially toxic elements in arable soils must not exceed certain prudent limits within the normal depth of cultivation as a result of sludge application. No sludge should be applied at any site where the soil concentration of any of the parameters mentioned, with the exception of molybdenum, exceed these limits. Maximum permissible concentrations of the potentially toxic elements in soil after application of sewage sludge (according to the UK Code of Practice) are given in Tablebelow. For zinc, copper and nickel, the maximum permissible concentrations vary with the pH of the soil because it is known that crop damage from phytotoxic elements is more likely to occur on acid soils. This Table also gives the maximum permissible average annual rates of addition of potentially toxic elements over a 10-year period.

Table: *Maximum Permissible Concentrations Of Potentially Toxic Elements In Soil After Application Of Sewage Sludge And Maximum Annual Rates Of Addition*

Potentially toxic element (PTE)	***Maximum permissible concentration of PTE in soil (mg/kg dry solids)***				***Maximum permissible average annual rate of PTE addition over a 10 year period (kg/ha)***[3]
	PH[1] 5.0 <5.5	pH[1] 5.5<6.0	pH 6.0-7.0	PH[2] > 7.0	
Zinc	200	250	300	450	15
Copper	80	100	135	200	7.5
Nickel	50	60	75	110	3
Cadmium	3[5]				0.15
Lead	300				15
Mercury	1				0.1
Chromium	400 (prov.)				15 (provisional)
*Molybdenum[4]	4				0.2
*Selenium	3				0.15
*Arsenic	50				0.7
*Fluoride	500				20

* These parameters are not subject to the provisions of Directive 86/278/ EEC.

[1] For soils of pH in the ranges of 5.0 < 5.5 and 5.5 < 6.0 the permitted concentrations of zinc, copper, nickel and cadmium are provisional and will be reviewed when current research into their effects on certain crops and livestock is completed.

[2] The increased permissible PTE concentrations in soils of pH greater than 7.0 apply only to soils containing more than 5 % calcium carbonate.

[3] The annual rate of application of PTE shall be determined by averaging over the 10-year period ending with the year of calculation.

[4] The accepted safe level of molybdenum in agricultural soils is 4 mg/kg. However, there are some areas in the UK where, for geological reasons, the natural concentration of this element in the soil exceeds this level. In such cases there may be no additional problems as a result of applying sludge, but this should not be done except in accordance with expert advice. This advice will take account of existing soil molybdenum levels and current arrangements to provide copper supplements to livestock.

[5] For pH 5.0 and above

Source: *Department of the Environment (1989)*

When sludge is applied to the surface of grassland, the concentrations of potentially toxic elements should be determined in soil samples taken to a depth of 7.5 cm. The maximum concentrations of these parameters should not exceed the limits set out in Table below. In order to minimize injestion of lead, cadmium and fluoride by livestock, the addition of these elements through sludge application to the surface should not exceed 3 times the 10 year average annual rates specified in Table above. Sludge to be surface applied to grassland should not contain lead or fluoride individually in excess of 1200 and 1000 mg/kg dry solids, respectively.

Table: *Maximum Permissible Concentrations Of Potentially Toxic Elements In Soil Under Grass After Application Of Sewage Sludge When Samples Taken To A Depth Of 7.5 Cm*

Potentially toxic element (PTE)	*Maximum permissible concentration of PTE in soil (mg/kg dry solids)*			
	pH 5.0 <5.5	pH 5.5<6.0	pH 6.0<7.0	PH[3] > 7.0
Zinc[1]	330	420	500	750
Copper[1]	130	170	225	330
Nickel[1]	80	100	125	180
Cadmium[2]	3/5[5]			
Lead	300			
Mercury	1.5			
Chromium	600 (prov.)			
*Molybdenum[4]	4			
*Selenium	5			
*Arsenic	50			
*Fluoride	500			

* These parameters are not subject to the provisions of Directive 86/278/EEC.

[1] The permitted concentrations of these elements will be subject to review when current research into their effects on the quality of grassland is completed. Until then, in cases where there is doubt about the practicality of ploughing or otherwise cultivating grassland, no sludge applications which would cause these concentrations to exceed the permitted levels specified should be made in accordance with specialist agricultural advice.

[2] The permitted concentration of cadmium will be subject to review when current research into its effect on grazing animals is completed. Until then, the concentration of this element may be raised to the permitted upper limit of 5 mg/kg as a result of sludge applications only under grass which is managed in rotation with arable crops and grown only for conservation. In all cases where grazing is permitted no sludge applications which would cause the concentration of cadmium to exceed the lower limit of 3 mg/kg shall be made.

Effects of Sludge on Soils and Crops

The natural background concentration of metals in the soil is normally less available for crop uptake and hence less hazardous than metals introduced through sewage sludge applications (Scheltinga, 1987). Research carried out in the U.K. (Carlton-Smith, 1987) has shown that the amounts of Cd, Ni, Cu, Zn and Pb applied in liquid sludge at three experimental sites could be accounted for by soil profile analyses five years after sludge applications, with the exception of Cu and Zn applied to a calcareous loam soil. These field experiments also determined the extent of transfer of metals from sludge-treated soil into the leaves and edible parts of six crops of major importance to UK agriculture and the effect of metals on yields of these crops.

Although all the plots received sufficient inorganic fertilizer to meet crop requirements for nutrients, the applications of sludge had some effects on crop yields. In 60% of the cases studied crop yields were not significantly affected but in 26% of the cases liquid sludge application resulted in significantly increased crop yields, attributed to the beneficial effects on soil structure. Reductions in wheat grain yield, from 6 - 10%, were noted on the clay and calcareous loam soils treated with liquid sludge and the sandy loam and clay soils treated with bed-dried sludge. However, this yield reduction was not thought to be due to metals but the most likely explanation was lodging of the crop as a result of excessive nitrogen in the soil.

Increases in metal concentrations in the soil due to sludge applications produced significant increases in Cd, Ni, Cu and Zn concentrations in the edible portion of most of the crops grown: wheat, potato, lettuce, red beet, cabbage and ryegrass. In most cases there was no significant increase of Pb in crop tissue in relation to Pb in the soil from sludge application, suggesting that lead is relatively unavailable to crops from the soil. The availability of metals to crops was found to be lower in soil treated with bed-dried sludge cake compared with liquid sludge, the extent being dependent on the crop. Even though the Ni, Cu and Zn concentrations in the soils treated with high rates of application of liquid and bed-dried sludges were close to the maximum levels set out in the EC Directive and the zinc equivalent of sludge addition exceeded the maximum permitted in U.K. guidelines, no phytotoxic effects of metals were evident, with one exception. This was in lettuce grown on clay soil, when Cu and Zn levels exceeded upper critical concentrations at high rates of sludge application.

Planting, Grazing and Harvesting Constraints

To minimize the potential risk to the health of humans, animals and plants it is necessary to coordinate sludge applications in time with planting, grazing or harvesting operations. Sludge must not be applied to growing soft fruit or vegetable crops nor used where crops are grown under permanent glass or plastic structures (Department of the Environment, 1989). The EC Directive (Council of the European Communities, 1986) requires a mandatory 3-week no grazing period for treated sludge applied to grassland but prohibits the spreading of untreated sludge on grassland unless injected. Treated sludge can be applied to growing cereal crops without constraint but should not be applied to growing turf within 3 months of harvesting or to fruit trees within 10 months of harvesting. When treated sludge is applied before planting such crops as cereals, grass, fodder, sugar beet, fruit trees, etc., no constraints apply but in the case of soft fruit and vegetables, the treated sludge should not be applied within 10 months of crop harvesting. In general, untreated sludge should only be cultivated or injected into the soil before planting crops but can be injected into growing grass or turf, with the constraints on minimum time to harvesting as already mentioned.

Environmental Protection

Care should always be taken when applying sewage sludge to land to prevent any form of adverse environmental impact. The sludge must not contain non-degradable materials, such as plastics, which would make land disposal unsightly. Movement of sludge by tanker from sewage treatment plant to agricultural land can create traffic problems and give rise to noise and odour nuisance. Vehicles should be carefully selected for their local suitability and routes chosen so as to minimize inconvenience to the public. Access to fields should be selected after consultation with the highway authority and special care must be taken to prevent vehicles carrying mud onto the highway.

Odour control is the most important environmental dimension of sludge application to land. Enclosed tankers should be used for transporting treated sludge, which tends to be less odorous than raw sludge. Discharge points for sludge from tankers or irrigators should be as near to the ground as is practicable and the liquid sludge trajectory should be kept low so as to minimize spray drift and visual impact. Untreated sludge should be injected under the soil surface using special vehicles or tankers fitted with injection equipment.

Great care is needed to prevent sludge running off onto roads or adjacent land, depending on topography, soil and weather conditions. On sloping land there is the risk of such runoff reaching watercourses and causing serious water pollution. Sludge application rates must be adjusted accordingly and, under certain circumstances, spreading might have to be discontinued. In addition to the problem of surface runoff, pollution may arise from the percolation of liquid sludge into land drains, particularly when injection techniques are used or liquid sludge is applied to dry fissured soils. In highly sensitive water pollution areas, sludge should be used only in accordance with the requirements of the pollution control authority as well as of good farming practice. Sludge storage on farms can optimize the transport and application operations but every effort must be made to ensure that storage facilities are secure.

Nematode

The nematodes or roundworms comprise the phylum Nematoda. They are a diverse animal phylum inhabiting a very broad range of environments. Nematode species can be difficult to distinguish; and although over 28,000 have been described, of which over 16,000 are parasitic, the total number of nematode species has been estimated to be about 1 million. Unlike cnidarians and flatworms, nematodes have tubular digestive systems with openings at both ends.

Habitats

Nematodes have successfully adapted to nearly every ecosystem from marine to fresh water, to soils, and from the polar regions to the tropics, as well as the highest to the lowest of elevations. They are ubiquitous in freshwater, marine, and terrestrial environments, where they often outnumber other animals in both individual and species counts, and are found in locations as diverse as mountains, deserts and oceanic trenches. They are found in every part of the earth's lithosphere. They represent, for example, 90% of all life forms on the ocean floor. Their numerical dominance, often exceeding a million individuals per square meter and accounting for about 80% of all individual animals on earth, their diversity of life cycles, and their presence at various trophic levels point at an important role in many ecosystems. Their many parasitic forms include pathogens in most plants and animals (including humans). Some nematodes can undergo cryptobiosis.

One group of carnivorous fungi, the nematophagous fungi, are predators of soil nematodes. They set enticements for the nematodes in the form of lassos or adhesive structures. Nematodes have even been found at great depth (0.9–3.6 km) below the surface of the Earth in gold mines in South Africa.

Nathan Cobb Described the Ubiquity of Nematodes on Earth Thus

In short, if all the matter in the universe except the nematodes were swept away, our world would still be dimly recognizable, and if, as disembodied spirits, we could then investigate it, we should find its mountains, hills, vales, rivers, lakes, and oceans represented by a film of nematodes. The location of towns would be decipherable, since for every massing of human beings there would be a corresponding massing of certain nematodes. Trees would still stand in ghostly rows representing our streets and highways. The location of the various plants and animals would still be decipherable, and, had we sufficient knowledge, in many cases even their species could be determined by an examination of their erstwhile nematode parasites."

Taxonomy and Systematics

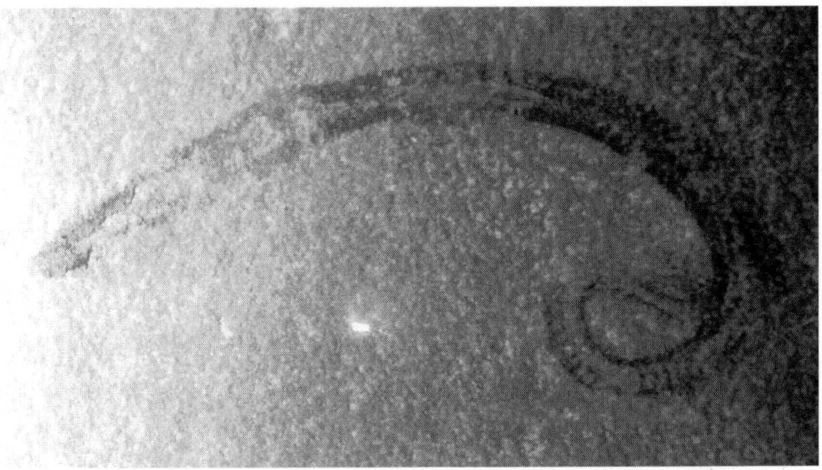

Figure: Eophasma jurasicum, *a fossilized nematode*

The group was originally defined by Karl Rudolphi in 1808 under the name Nematoidea, from Ancient Greek *nêma, nêmatos*, 'thread' and 'species'. It was reclassified as family Nematodes by Burmeister in 1837 and order Nematoda by K. M. Diesing in 1861.

At its origin, the "Nematoidea" included both roundworms and horsehair worms. Along with Acanthocephala, Trematoda and Cestoidea, it formed the group Entozoa. The first differentiation of roundworms from horsehair worms, though erroneous, is due to von Siebold (1843) with orders Nematoidea and Gordiacei (Gordiacea). They were classed along with Acanthocephala in the new phylum Nemathelminthes (today obsolete) by Gegenbaur (1859). The taxon Nematoidea, including the family Gordiidae (horsehair worms), was then promoted to the rank of phylum by Ray Lankester (1877). In 1919, Nathan Cobb proposed that roundworms should be recognized alone as a phylum. He argued they should be called nema(s) in English rather than "nematodes" and defined the taxon Nemates (Latin plural of *nema*). Since Cobb was the first to exclude all but nematodes from the group, some sources consider the valid taxon name to be Nemates or Nemata, rather than Nematoda.

Phylogeny

The relationships of the nematodes and their close relatives among the protostomian Metazoa are unresolved. Traditionally, they were held to be a lineage of their own but in the 1990s, they were proposed to form the group Ecdysozoa together with moulting animals, such as arthropods. The identity of the closest living relatives of the Nematoda has always been considered to be well resolved. Morphological characters and molecular phylogenies agree with placement of the roundworms as a sister taxon to the parasitic horsehair worms (Nematomorpha); together they make up the Nematoida. Together with the Scalidophora (formerly Cephalorhyncha), the Nematoida form the Introverta. It is entirely unclear whether the Introverta are, in turn, the closest living relatives of the enigmatic Gastrotricha; if so, they are considered a clade Cycloneuralia, but much disagreement occurs both between and among the available morphological and molecular data. The Cycloneuralia or the Introverta—depending on the validity of the former—are often ranked as a superphylum.

Nematode Systematics

Due to the lack of knowledge regarding many nematodes, their systematics is contentious. An earliest and influential classification was proposed by Chitwood and Chitwood—later revised by Chitwood—who divided the phylum into two—the Aphasmidia and the Phasmidia. These were later renamed Adenophorea (gland bearers) and Secernentea (secretors), respectively. The Secernentea share several characteristics, including the presence of phasmids, a pair of sensory organs located in the lateral posterior region, and this was used as the basis for this division. This scheme was adhered to in many later classifications, though the Adenophorea were not a uniform group.

Initial DNA sequence studies suggested the existence of five clades:

- Dorylaimia
- Enoplia
- Spirurina
- Tylenchina
- Rhabditina

As it seems, the Secernentea are indeed a natural group of closest relatives. But the "Adenophorea" appear to be a paraphyletic assemblage of roundworms simply retaining a good number of ancestral traits. The old Enoplia do not seem to be monophyletic either, but to contain two distinct lineages. The old group "Chromadoria" seem to be another paraphyletic assemblage, with the Monhysterida representing a very ancient minor group of nematodes. Among the Secernentea, the Diplogasteria may need to be united with the Rhabditia, while the Tylenchia might be paraphyletic with the Rhabditia.

The understanding of roundworm systematics and phylogeny as of 2002 is summarised below:

Phylum Nematoda

- Basal order Monhysterida
- Class Dorylaimea
- Class Enoplea
- Class Secernentea
  - o Subclass Diplogasteria (disputed)
  - o Subclass Rhabditia (paraphyletic?)
  - o Subclass Spiruria
  - o Subclass Tylenchia (disputed)
- "Chromadorea" assemblage

Later work has suggested the presence of 12 clades. The Secernentea—a group that includes virtually all major animal and plant 'nematode' parasites—apparently arose from within the Adenophorea.

A major effort to improve the systematics of this phylum is in progress and being organised by the 959 Nematode Genomes.

A complete checklist of the World's nematode species can be found in the World Species Index:Nematoda.

An analysis of the mitochondrial DNA suggests that the following groupings are valid

- subclass Dorylaimia
- orders Rhabditida, Trichinellida and Mermithida
- suborder Rhabditina
- infraorders Spiruromorpha and Oxyuridomorpha

The Ascaridomorpha, Rhabditomorpha and Diplogasteromorpha appear to be related.

The suborders Spirurina and Tylenchina and the infraorders Rhabditomorpha, Panagrolaimomorpha and Tylenchomorpha are paraphytic.

The monophyly of the Ascaridomorph is uncertain.

Anatomy

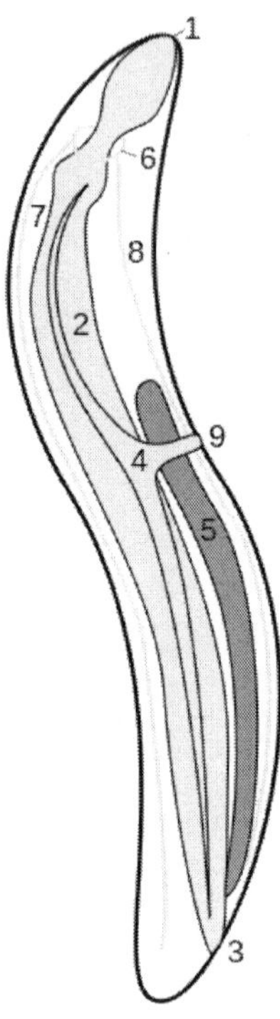

Figure: *Schematic drawing of the anatomy of a male nematode. 1 mouth opening, 2 intestine, 3 cloacal opening, 4 organ of excretion, 5 testis, 6 circumpharyngeal ring of nervous system, 7 dorsal trunk of nervous system, 8 ventral trunk of nervous system, 9 excretion pore*

Nematodes are slender worms, typically less than 2.5 mm (0.098 in) long. The smallest nematodes are microscopic, while free-living species can reach as much as 5 cm (2.0 in), and some parasitic species are larger still, reaching over a centre in length. The body is often ornamented with ridges, rings, bristles, or other distinctive structures.

The head of a nematode is relatively distinct. Whereas the rest of the body is bilaterally symmetrical, the head is radially symmetrical, with sensory bristles and, in many cases, solid 'head-shields' radiating outwards around the mouth. The mouth has either three or six lips, which often bear a series of teeth on their inner edges. An adhesive 'caudal gland' is often found at the tip of the tail.

The epidermis is either a syncytium or a single layer of cells, and is covered by a thick collagenous cuticle. The cuticle is often of complex structure, and may have two or three distinct layers. Underneath the epidermis lies a layer of longitudinal muscle cells. The relatively rigid cuticle works with the muscles to create a hydroskeleton as nematodes lack circumferential muscles. Projections run from the inner surface of muscle cells towards the nerve cords; this is a unique arrangement in the animal kingdom, in which nerve cells normally extend fibres into the muscles rather than *vice versa*.

Digestive System

The oral cavity is lined with cuticle, which is often strengthened with ridges or other structures, and, especially in carnivorous species, may bear a number of teeth. The mouth often includes a sharp stylet, which the animal can thrust into its prey. In some species, the stylet is hollow, and can be used to suck liquids from plants or animals.

The oral cavity opens into a muscular, sucking pharynx, also lined with cuticle. Digestive glands are found in this region of the gut, producing enzymes that start to break down the food. In stylet-bearing species, these may even be injected into the prey.

There is no stomach, with the pharynx connecting directly to a muscleless intestine that forms the main length of the gut. This produces further enzymes, and also absorbs nutrients through its single cell thick lining. The last portion of the intestine is lined by cuticle, forming a rectum, which expels waste through the anus just below and in front of the tip of the tail. Movement of food through the digestive system is the result of body movements of the worm. The intestine has valves or sphincters at either end to help control the movement of food through the body.

Excretory System

Nitrogenous waste is excreted in the form of ammonia through the body wall, and is not associated with any specific organs. However, the structures for excreting salt to maintain osmoregulation are typically more complex.

In many marine nematodes, one or two unicellular 'renette glands' excrete salt through a pore on the underside of the animal, close to the pharynx. In most other nematodes, these specialised cells have been replaced by an organ consisting of two parallel ducts connected by a single transverse duct. This transverse duct opens into a common canal that runs to the excretory pore.

Nervous System

Four peripheral nerves run the length of the body on the dorsal, ventral, and lateral surfaces. Each nerve lies within a cord of connective tissue lying beneath the cuticle and between the muscle cells. The ventral nerve is the largest, and has a double structure forward of the excretory pore. The dorsal nerve is responsible for motor control, while the lateral nerves are sensory, and the ventral combines both functions.

At the anterior end of the animal, the nerves branch from a dense, circular nerve ring surrounding the pharynx, and serving as the brain. Smaller nerves run forward from the ring to supply the sensory organs of the head.

The bodies of nematodes are covered in numerous sensory bristles and papillae that together provide a sense of touch. Behind the sensory bristles on the head lie two small pits, or 'amphids'. These are well supplied with nerve cells, and are probably chemoreception organs. A few aquatic nematodes possess what appear to be pigmented eye-spots, but is unclear whether or not these are actually sensory in nature.

Reproduction

Most nematode species are dioecious, with separate male and female individuals. Both sexes possess one or two tubular gonads. In males, the sperm are produced at the end of the gonad, and migrate along its length as they mature. The testes each open into a relatively wide sperm duct and then into a glandular and muscular ejaculatory duct associated with the cloaca. In females, the ovaries each open into an oviduct and then a glandular uterus. The uteri both open into a common vagina, usually located in the middle of the ventral surface.

Reproduction is usually sexual. Males are usually smaller than females (often much smaller) and often have a characteristically bent tail for holding the female. During copulation, one or more chitinized spicules move out of the cloaca and are inserted into genital pore of the female. Amoeboid sperm crawl along the spicule into the female worm. Nematode sperm is thought to be the only eukaryotic cell without the globular protein G-actin.

Eggs may be embryonated or unembryonated when passed by the female, meaning their fertilized eggs may not yet be developed. A few species are known to be ovoviviparous. The eggs are protected by an outer shell, secreted by the uterus. In free-living roundworms, the eggs hatch into larvae, which appear essentially identical to the adults, except for an underdeveloped reproductive system; in parasitic roundworms, the life cycle is often much more complicated.

Nematodes as a whole possess a wide range of modes of reproduction. Some nematodes, such as *Heterorhabditis* spp., undergo a process called *endotokia matricida*: intrauterine birth causing maternal death. Some nematodes are hermaphroditic, and keep their self-fertilized eggs inside the uterus until they hatch. The juvenile nematodes will then ingest the parent nematode. This process is significantly promoted in environments with a low food supply.

The nematode model species *Caenorhabditis elegans* and *C. briggsae* exhibit androdioecy, which is very rare among animals. The single genus *Meloidogyne* (root-knot nematodes) exhibit a range of reproductive modes, including sexual reproduction, facultative sexuality (in which most, but not all, generations reproduce asexually), and both meiotic and mitotic parthenogenesis.

The genus *Mesorhabditis* exhibits an unusual form of parthenogenesis, in which sperm-producing males copulate with females, but the sperm do not fuse with the ovum. Contact with the sperm is essential for the ovum to begin dividing, but because there is no fusion of the cells, the male contributes no genetic material to the offspring, which are essentially clones of the female.

Free-living Species

In free-living species, development usually consists of four molts of the cuticle during growth. Different species feed on materials as varied as algae, fungi, small animals, fecal matter, dead organisms and living tissues. Free-living marine nematodes are important and abundant members of the meiobenthos. They play an important role in the decomposition process, aid in recycling of nutrients in marine environments, and are sensitive to changes in the environment caused by pollution. One roundworm of note, *Caenorhabditis elegans*, lives in the soil and has found much use as a model organism. *C. elegans* has had its entire genome sequenced, as well as the developmental fate of every cell determined, and every neuron mapped.

Parasitic Species

Nematodes commonly parasitic on humans include ascarids (*Ascaris*), filarias, hookworms, pinworms (*Enterobius*) and whipworms (*Trichuris trichiura*). The species *Trichinella spiralis*, commonly known as the 'trichina worm', occurs in rats, pigs, and humans, and is responsible for the disease trichinosis. *Baylisascaris* usually infests wild animals, but can be deadly to humans, as well. *Dirofilaria immitis* heartworms are known for causing heartworm disease by inhabiting the hearts, arteries, and lungs of dogs and some cats. *Haemonchus contortus* is one of the most abundant infectious agents in sheep around the world, causing great economic damage to sheep. In contrast, entomopathogenic nematodes parasitize insects and are considered by humans to be beneficial.

One form of nematode is entirely dependent upon fig wasps, which are the sole source of fig fertilization. They prey upon the wasps, riding them from the ripe fig of the wasp's birth to the fig flower of its death, where they kill the wasp, and their offspring await the birth of the next generation of wasps as the fig ripens.

A newly discovered parasitic tetradonematid nematode, *Myrmeconema neotropicum*, apparently induces fruit mimicry in the tropical ant *Cephalotes atratus*. Infected ants develop bright red gasters, tend to be more sluggish, and walk with their gasters in a conspicuous elevated position. These changes likely cause frugivorous birds to confuse the infected ants for berries, and eat them. Parasite eggs passed in the bird's feces are subsequently collected by foraging *Cephalotes atratus* and are fed to their larvae, thus completing the life cycle of *M. neotropicum*.

Plant-parasitic nematodes include several groups causing severe crop losses. The most common genera are *Aphelenchoides* (foliar nematodes), *Ditylenchus*, *Globodera* (potato cyst nematodes), *Heterodera* (soybean cyst nematodes), *Longidorus*, *Meloidogyne* (root-knot nematodes), *Nacobbus*, *Pratylenchus* (lesion nematodes), *Trichodorus* and *Xiphinema* (dagger nematodes). Several phytoparasitic nematode species cause histological damages to roots, including the formation of visible galls (e.g. by root-knot nematodes), which are useful characters for their diagnostic in the field. Some nematode species transmit plant viruses through their feeding activity on roots. One of them is *Xiphinema index*, vector of grapevine fanleaf virus), an important disease of grapes, another one is *Xiphinema diversicaudatum*, vector of arabis mosaic virus.

Other nematodes attack bark and forest trees. The most important representative of this group is *Bursaphelenchus xylophilus*, the pine wood nematode, present in Asia and America and recently discovered in Europe.

Agriculture and Horticulture

Depending on the species, a nematode may be beneficial or detrimental to plant health. From agricultural and horticulture perspectives, the two categories of nematodes are the predatory ones, which will kill garden pests like cutworms, and the pest nematodes, like the root-knot nematode, which attack plants, and those that act as vectors spreading plant viruses between crop plants. Predatory nematodes can be bred by soaking a specific recipe of leaves and other detritus in water, in a dark, cool place, and can even be purchased as an organic form of pest control.

Rotations of plants with nematode-resistant species or varieties is one means of managing parasitic nematode infestations. For example, marigolds, grown over one or more seasons (the effect is cumulative), can be used to control nematodes. Another is treatment with natural antagonists such as the fungus *Gliocladium roseum*. Chitosan, a natural biocontrol, elicits plant defence responses to destroy parasitic cyst nematodes on roots of soybean, corn, sugar beet, potato and tomato crops without harming beneficial nematodes in the soil. Furthermore, soil steaming is an efficient method to kill nematodes before planting the crop, but indiscriminately eliminates both harmful and beneficial ones.

The Golden Nematode is a particularly harmful variety of nematode pest that has resulted in quarantines and crop failures worldwide.

CSIRO has found a 13- to 14-fold reduction of nematode population densities in plots having Indian mustard (*Brassica juncea*) green manure or seed meal in the soil.

Hundreds of *Caenorhabditis elegans* were featured in a research project on NASA's STS-107 space mission, and were the only known living organisms to have survived the Space Shuttle Columbia disaster.

Cutworm

Cutworms are not worms, biologically speaking, but caterpillars; they are moth larvae that hide under litter or soil during the day, coming out in the dark to feed on plants. A larva typically attacks the first part of the plant it encounters, namely the stem, often of a seedling, and consequently cuts it down; hence the name *cutworm*.

Feeding and Hiding

Cutworm larvae vary in their feeding behaviour; some remain with the plant they cut down and feed on it, while others often move on after eating a small amount from a felled seedling; such a wasteful mode of feeding results in disproportionate damage to crops. Cutworms accordingly are serious pests to gardeners in general, but to vegetable and grain farmers in particular. For example, it has been suggested that in South Africa for one, *Agrotis segetum* is the second worst pest of maize.

Note that the cutworm mode of feeding is only one version of a strategy of avoiding predators and parasitoids by day. Many other caterpillars, including Noctuidae and some kinds of processionary caterpillars, come out at night to feed, but hide again as soon as the sky begins to grow lighter. Some, for example *Klugeana philoxalis* attack low-growing forbs such as *Oxalis* in the dark, and drop to the ground as soon as a light is flashed on them. Others will climb trees such as species of Acacia nightly, leaving trails of silk, but they leave individual trails, not common trails like processionary caterpillars. The fruit-piercing moth *Serrodes partita* similarly lives under litter beneath its food plant, the tree *Pappea capensis*

Species and Habits

The term cutworm applies mainly to larvae of various species in the Noctuidae, a large family of moths; however many Noctuid species are not cutworms, and some moths whose larvae have essentially the same habit, which justifies calling them cutworms, are not Noctuids. The larvae of the Turnip moth, (Agrotis segetum, Agrotis ipsilon, Agrotis exclamationis), are well-known Noctuids whose larvae are very damaging cutworms. Cutworms are notorious agricultural and garden pests. They are voracious leaf, bud, and stem feeders and can destroy entire plants. They get their name from their habit of "cutting" off a seedling at ground level by chewing through the stem. Some species are subterranean and eat roots.

Appearance and Control

Cutworms are usually green, brown, grey, or yellow soft-bodied caterpillars, often with longitudinal stripes, up to 1 inch (2.5 centimeters) in length. There are many variations among the genera. There also are variations in their biology and control, so the following extension material must be applied only as appropriate to the region.

In many climates, cutworms will winter under the soil, either as final instar larvae or as pupae. This affords farmers an opportunity for control. Winter ploughing will kill many of the pests, and expose many more to predators. In suitable areas this is a powerful means of control, for example in grain fields. The same principle permits some domestic gardeners to kill the caterpillars without the problems associated with the use of pesticides; the first line of control can be to till the soil some weeks before planting to destroy any dormant larvae. Also, at any time during the season, if the population has been reasonably well controlled, but there are signs of localised cutworm attack, the domestic gardener may be able to deal with the problem simply by digging the soil and wet foliage to about 2 inches deep, and killing the caterpillars manually.

Starvation also can be effective when it is practical to keep weeds down before the growing season, by systematic cultivation. Together with reducing manure and compost, relying instead on other forms of fertilizer, this can improve control by discouraging cutworm moths from laying their eggs, and depriving the larvae of food.

Baits also can be effective where starvation strategies have been applied reasonably successfully; a sweetened bran mash containing a suitable stomach poison can be very effective against the small numbers of surviving caterpillars. The mash should be too crumbly and too thinly scattered to leave any lumps on the ground that domestic animals or desirable wild animals might otherwise pick up.

Because cutworms attack the first part of the plant they find at night, plant collars made of aluminum, or even cardboard barriers, can offer effective protection.

Hosts for Parasitoids

Cutworms are hosts for numerous parasitoid wasps and flies, including species of Braconidae, Ichneumonidae, Tachinidae and Eulophidae, with rates of parasitism as high as 75 - 80%.

Root-knot Nematode

Root-knot nematodes are plant-parasitic nematodes from the genus *Meloidogyne.* They exist in soil in areas with hot climates or short winters. About 2000 plants are susceptible to infection by root-knot nematodes and they cause approximately 5% of global crop loss. Root-knot nematode larvae infect plant roots, causing the development of root-knot galls that drain the plant's photosynthate and nutrients. Infection of young plants may be lethal, while infection of mature plants causes decreased yield.

Economic Impact

Root-knot nematodes (*Meloidogyne* spp.) are one of the three most economically damaging genera of plant-parasitic nematodes on horticultural and field crops. Root-knot nematodes are distributed worldwide, and are obligate parasites of the roots of thousands of plant species, including monocotyledonous and dicotyledonous, herbaceous and woody plants. The genus includes more than 60 species, with some species having several races. Four *Meloidogyne* species (*M. javanica, M. arenaria, M. incognita, and M. hapla*) are major pests worldwide, with another seven being important on a local basis. *Meloidogyne* occurs in 23 of 43 crops listed as having plant-parasitic nematodes of major importance, ranging from field crops, through pasture and grasses, to horticultural, ornamental and vegetable crops. If root-knot nematodes become established in deep-rooted, perennial crops, control is difficult and options are limited. *Meloidogyne* spp. were first reported in cassava by Neal in 1889.

Figure: *Swiss chard with root-knot nematode damage*

Damage on cassava is variable depending on cultivar planted, and ranges from negligible to seriously damaging. Early-season infection leads to worse damage. In most crops, nematode damage reduces plant health and growth; in cassava, though, nematode damage sometimes leads to increased aerial growth as the plants try to compensate. This possibly enables the plant to maintain a reasonable level of production. Therefore, aerial correlations to nematode density can be positive, negative or not at all. Vegetable crops grown in warm climates can experience severe losses from root-knot nematodes, and are often routinely treated with a chemical nematicide. Root-knot nematode damage results in poor growth, a decline in quality and yield of the crop and reduced resistance to other stresses (e.g. drought, other diseases). A high level of damage can lead to total crop loss. Nematode-damaged roots do not use water and fertilizers as effectively, leading to additional losses for the grower. In cassava, it has been suggested that levels of *Meloidogyne* spp. that are sufficient to cause injury rarely occur naturally. However, with changing farming systems, in a disease complex or weakened by other factors, nematode damage is likely to be associated with other problems.

Control

Root-knot nematodes (*Meloidogyne* spp.) can be controlled with a biocontrol agent *Paecilomyces lilacinus, Pasteuria pentrans. Juglone* effects on Development of the plant parasitic nematode *(Meloidogyne Spp.)* on *Arachis hypogaea L.* Napthoquinones controlling the root-knot nematode *Meloidogyne spp.* Zinc (II) complexes of NNO Tri-Dentate Heterocyclic Schiff bases controlled pytonematodes. Microwave assisted synthesis,substituted 2-Aminothiazoles also controlled plant nematodes. Thiazole Schiff Bases controlled pytonematodes.

Life Cycle

All nematodes pass through an embryonic stage, four juvenile stages (J1–J4) and an adult stage. Juvenile *Meloidogynes* parasites hatch from eggs as vermiform, second-stage juveniles (J2), the first moult having occurred within the egg. Newly-hatched juveniles have a short free-living stage in the soil, in the rhizosphere of the host plants. They may reinvade the host plants of their parent or migrate through the soil to find a new host root. J2 larvae do not feed during the free-living stage, but use lipids stored in the gut.

An excellent model system for the study of the parasitic behaviour of plant-parasitic nematodes has been developed using *Arabidopsis thaliana* as a model host. The *Arabidopsis* roots are initially small and transparent, enabling every detail to be seen. Invasion and migration in the root was studied using *M. incognita*. Briefly, second stage juveniles invade in the root elongation region and migrate in the root until they became sedentary. Signals from the J2 promote parenchyma cells near the head of the J2 to become multinucleate to form feeding cells, generally known as giant cells, from which the J2 and later the adults feed. Concomitant with giant cell formation, the surrounding root tissue gives rise to a gall in which the developing juvenile is embedded. Juveniles first feed from the giant cells about 24 hours after becoming sedentary.

After further feeding, the J2s undergo morphological changes and become saccate. Without further feeding, they moult three times and eventually become adults. In females, which are close to spherical, feeding resumes and the reproductive system develops. The life span of an adult female may extend to three months, and many hundreds of eggs can be produced. Females can continue egg laying after harvest of aerial parts of the plant and the survival stage between crops is generally within the egg.

The length of the life cycle is temperature-dependent. The relationship between rate of development and temperature is linear over much of the root-knot nematode life cycle, though it is possible the component stages of the life cycle, e.g. egg development, host root invasion or growth, have slightly different optima. Species within the *Meloidogyne* genus also have different temperature optima. In *M. javanica*, development occurs between 13 and 34°C, with optimal development at about 29°C.

Gelatinous Matrix

Root-knot nematode females lay eggs into a gelatinous matrix (GM), which is produced by six rectal glands and secreted before and during egg laying. The matrix initially forms a canal through the outer layers of root tissue and later surrounds the eggs, providing a barrier to water loss by maintaining a high moisture level around the eggs. As the gelatinous matrix ages, it becomes tanned, turning from a sticky, colourless jelly to an orange-brown substance which appears layered.

Egg Formation and Development

Egg formation in *M. javanica* has been studied in detail, and is similar to egg formation in the well studied, free-living nematode *Caenorhabditis elegans*. Embryogenesis has also been studied, and the stages of development are easily identifiable with a phase contrast microscope following preparation of an egg mass squash. The egg is formed as one cell, with two-cell, four-cell and eight-cell stages recognisable. Further cell division leads to the tadpole stage, with further elongation resulting in the first stage juvenile, which is roughly four times as long as the egg. The J1 stage of *C. elegans* has 558 cells, and the J1 of *M. javanica* likely has a similar number, since all nematodes are morphologically and anatomically similar. The egg shell has three layers, with the vitelline layer outermost, then a chitinous layer and a lipid layer innermost.

Egg Hatching

Preceded by induced changes in eggshell permeability, hatching may involve physical and/or enzymatic processes in plant-parasitic nematodes. Cyst nematodes, such as *Globodera rostochiensis*, may require a specific signal from the root exudates of the host to trigger hatching. Root-knot nematodes are generally unaffected by the presence of a host, but hatch freely at the appropriate temperature when water is available. However, in an egg mass or cyst, not all eggs will hatch when the conditions are optimal for their particular species, leaving some eggs to hatch at a later date. Ammonium ions have been shown to inhibit hatching and to reduce the plant-penetration ability of *M. incognita* juveniles that do hatch.

Reproduction

Root-knot nematodes exhibit a range of reproductive modes, including sexuality (amphimixis), facultative sexuality, meiotic parthenogenesis (automixis) and mitotic parthenogenesis (apomixis).

Plant Virus

Figure: Pepper mild mottle virus

Plant viruses are viruses that affect plants. Like all other viruses, plant viruses are obligate intracellular parasites that do not have the molecular machinery to replicate without a host. Plant viruses are pathogenic to higher plants. While this article does not intend to list all plant viruses, it discusses some important viruses as well as their uses in plant molecular biology.

Overview

Although plant viruses are not nearly as well understood as the animal counterparts, one plant virus has become iconic. The first virus to be discovered was *Tobacco mosaic virus* (TMV). This and other viruses cause an estimated US$60 billion loss in crop yields worldwide each year. Plant viruses are grouped into 73 genera and 49 families.

To transmit from one plant to another and from one plant cell to another, plant viruses must use strategies that are usually different from animal viruses. Plants do not move, and so plant-to-plant transmission usually involves vectors (such as insects). Plant cells are surrounded by solid cell walls, therefore transport through plasmodesmata is the preferred path for virions to move between plant cells. Plants probably have specialized mechanisms for transporting mRNAs through plasmodesmata, and these mechanisms are thought to be used by RNA viruses to spread from one cell to another.

Plant defences against viral infection include, among other measures, the use of siRNA in response to dsRNA. Most plant viruses encode a protein to suppress this response. Plants also reduce transport through plasmodesmata in response to injury.

History

The discovery of plant viruses causing disease is often accredited to Martinus Beijerinck who determined, in 1898, that plant sap obtained from tobacco leaves with the "mosaic disease" remained infectious when passed through a porcelain filter. This was in contrast to bacteria microorganisms, which were retained by the filter. Beijerinck referred to the infectious filtrate as a "contagium vivum fluidum", thus the coinage of the modern term "virus".

After the initial discovery of the 'viral concept' there was need to classify any other known viral diseases based on the mode of transmission even though microscopic observation proved fruitless. In 1939 Holmes published a classification list of 129 plant viruses. This was expanded and in 1999 there were 977 officially recognized, and some provisional, plant virus species.

The purification (crystallization) of TMV was first performed by Wendell Stanley, who published his findings in 1935, although he did not determine that the RNA was the infectious material. However, he received the Nobel Prize in Chemistry in 1946. In the 1950s a discovery by two labs simultaneously proved that the purified RNA of the TMV was infectious which reinforced the argument. The RNA carries genetic information to code for the production of new infectious particles.

More recently virus research has been focused on understanding the genetics and molecular biology of plant virus genomes, with a particular interest in determining how the virus can replicate, move and infect plants. Understanding the virus genetics and protein functions has been used to explore the potential for commercial use by biotechnology companies. In particular, viral-derived sequences have been used to provide an understanding of novel forms of resistance. The recent boom in technology allowing humans to manipulate plant viruses may provide new strategies for production of value-added proteins in plants.

Structure

Viruses are extremely small and can only be observed with an electron microscope. The structure of a virus is given by its coat of proteins, which surround the viral genome. Assembly of viral particles takes place spontaneously.

Over 50% of known plant viruses are rod-shaped (flexuous or rigid). The length of the particle is normally dependent on the genome but it is usually between 300–500 nm with a diameter of 15–20 nm. Protein subunits can be placed around the circumference of a circle to form a disc. In the presence of the viral genome, the discs are stacked, then a tube is created with room for the nucleic acid genome in the middle.

The second most common structure amongst plant viruses are isometric particles. They are 40–50 nm in diameter. In cases when there is only a single coat protein, the basic structure consists of 60 T subunits, where T is an integer. Some viruses may have 2 coat proteins are the associate to form an icosahedral shaped particle.

There are three genera of *Geminiviridae* that possess geminate particles which are like two isometric particles stuck together.

A very small number of plant viruses have, in addition to their coat proteins, a lipid envelope. This is derived from the plant cell membrane as the virus particle buds off from the cell.

Transmission of Plant Viruses

Through Sap: Viruses can be spread by direct transfer of sap by contact of a wounded plant with a healthy one. Such contact may occur during agricultural practices, as by damage caused by tools or hands, or naturally, as by an animal feeding on the plant. Generally TMV, potato viruses and cucumber mosaic viruses are transmitted via sap.

Insects

Plant viruses need to be transmitted by a vector, most often insects such as leafhoppers. One class of viruses, the Rhabdoviridae, has been proposed to actually be insect viruses that have evolved to replicate in plants. The chosen insect vector of a plant virus will often be the determining factor in that virus's host range: it can only infect plants that the insect vector feeds upon. This was shown in part when the old world white fly made it to the USA, where it transferred many plant viruses into new hosts. Depending on the way they are transmitted, plant viruses are classified as non-persistent, semi-persistent and persistent. In non-persistent transmission, viruses become attached to the distal tip of the stylet of the insect and on the next plant it feeds on, it inoculates it with the virus. Semi-persistent viral transmission involves the virus entering the foregut of the insect. Those viruses that manage to pass through the gut into the haemolymph and then to the salivary glands are known as persistent. There are two sub-classes of persistent viruses: propagative and circulative. Propagative viruses are able to replicate in both the plant and the insect (and may have originally been insect viruses), whereas circulative can not. Circulative viruses are protected inside aphids by the chaperone protein symbionin produced by bacterial symbionts. Many plant viruses encode within their genome polypeptides with domains essential for transmission by insects. In non-persistent and semi-persistent viruses, these domains are in the coat protein and another protein known as the helper component. A bridging hypothesis has been proposed to explain how these proteins aid in insect-mediated viral transmission. The helper component will bind to the specific domain of the coat protein, and then the insect mouthparts — creating a bridge. In persistent propagative viruses, such as tomato spotted wilt virus (TSWV), there is often a lipid coat surrounding the proteins that is not seen in the other classes of plant viruses. In the case of TSWV, 2 viral proteins are expressed in this lipid envelope. It has been proposed that the viruses bind via these proteins and are then taken into the insect cell by receptor-mediated endocytosis.

Nematodes

Soil-borne nematodes also have been shown to transmit viruses. They acquire and transmit them by feeding on infected roots. Viruses can be transmitted both non-persistently and persistently, but there is no evidence of viruses being able to replicate in nematodes. The virions attach to the stylet (feeding organ) or to the gut when they feed on an infected plant and can then unattach during later feeding to infect other plants. Examples of viruses that can be transmitted by nematodes include tobacco ringspot virus and tobacco rattle virus.

Plasmodiophorids

A number of virus genera are transmitted, both persistently and non-persistently, by soil borne zoosporic protozoa. These protozoa are not phytopathogenic themselves, but parasitic. Transmission of the virus takes place when they become associated with the plant roots. Examples include *Polymyxa graminis*, which has been shown to transmit plant viral diseases in cereal crops and *Polymyxa betae* which transmits Beet necrotic yellow vein virus. Plasmodiophorids also create wounds in the plant's root through which other viruses can enter.

Seed and Pollen Borne Viruses

Plant virus transmission from generation to generation occurs in about 20% of plant viruses. When viruses are transmitted by seeds, the seed is infected in the generative cells and the virus is maintained in the germ cells and sometimes, but less often, in the seed coat. When the growth and development of plants is delayed because of situations like unfavourable weather, there is an increase in the amount of virus infections in seeds. There does not seem to be a correlation between the location of the seed on the plant and its chances of being infected. Little is known about the mechanisms involved in the transmission of plant viruses via seeds, although it is known that it is environmentally influenced and that seed transmission occurs because of a direct invasion of the embryo via the ovule or by an indirect route with an attack on the embryo mediated by infected gametes. These processes can occur concurrently or separately depending on the host plant. It is unknown how the virus is able to directly invade and cross the embryo and boundary between the parental and progeny generations in the ovule. Many plants species can be infected through seeds including but not limited to the families Leguminosae, Solanaceae, Compositae, Rosaceae, Curcurbitaceae, Gramineae. Bean common mosaic virus is transmitted through seeds.

Direct Plant-to-human Transmission

Researchers from the University of the Mediterranean in Marseille, France have found evidence that suggest a virus common to peppers, the Pepper Mild Mottle Virus (PMMoV) may have moved on to infect humans. This is a very rare and highly unlikely event as, to enter a cell and replicate, a virus must "bind to a receptor on its surface, and a plant virus would be highly unlikely to recognize a receptor on a human cell. One possibility is that the virus does not infect human cells directly. Instead, the naked viral RNA may alter the function of the cells through a mechanism similar to RNA interference, in which the presence of certain RNA sequences can turn genes on and off," according to Virologist Robert Garry from the Tulane University in New Orleans, Louisiana.

Symptoms include being more likely to have fever, abdominal pain, and itching.

Translation of Plant Viral Proteins

75% of plant viruses have genomes that consist of single stranded RNA (ssRNA). 65% of plant viruses have +ssRNA, meaning that they are in the same sense orientation as messenger RNA but 10% have -ssRNA, meaning they must be converted to +ssRNA before they can be translated. 5% are double stranded RNA and so can be immediately translated as +ssRNA viruses. 3% require a reverse transcriptase enzyme to convert between RNA and DNA. 17% of plant viruses are ssDNA and very few are dsDNA, in contrast a quarter of animal viruses are dsDNA and three quarters of bacteriophage are dsDNA. Viruses use the plant ribosomes to produce the 4-10 proteins encoded by their genome. However, since all of the proteins are encoded on a single strand (that is, they are polycistronic) this will mean that the ribosome will either only produce one protein, as it will terminate translation at the first stop codon, or that a polyprotein will be produced. Plant viruses have had to evolve special techniques to allow the production of viral proteins by plant cells.

Cap

For translation to occur, eukaryotic mRNAs require a 5' Cap structure. This means that viruses must also have one. This normally consists of 7MeGpppN where N is normally adenine or guanine. The viruses encode a protein, normally a replicase, with a methyltransferase activity to allow this.

Some viruses are cap-snatchers. During this process, a ^{7m}G-capped host mRNA is recruited by the viral transcriptase complex and subsequently cleaved by a virally encoded endonuclease. The resulting capped leader RNA is used to prime transcription on the viral genome.

However some plant viruses do not use cap, yet translate efficiently due to cap-independent translation enhancers present in 5' and 3' untranslated regions of viral mRNA.

Readthrough

Some viruses (e.g. tobacco mosaic virus (TMV)) have RNA sequences that contain a "leaky" stop codon. In TMV 95% of the time the host ribosome will terminate the synthesis of the polypeptide at this codon but the rest of the time it continues past it.

This means that 5% of the proteins produced are larger than and different from the others normally produced, which can be thought of as a crude form of transcriptional regulation. In TMV, this extra sequence of polypeptide is an RNA polymerase that replicates its genome.

Production of Sub-genomic RNAs

Some viruses use the production of subgenomic RNAs to ensure the translation of all proteins within their genomes. In this process the first protein encoded on the genome, and this the first to be translated, is a replicase.

This protein will act on the rest of the genome producing negative strand sub-genomic RNAs then act upon these to form positive strand sub-genomic RNAs that are essentially mRNAs ready for translation.

Segmented Genomes

Some viral families, such as the *Bromoviridae* instead opt to have multi-partite genomes, genomes split between multiple viral particles. For infection to occur, the plant must be infected with all particles across the genome. For instance *Brome mosaic virus* has a genome split between 3 viral particles, and all 3 particles with the different RNAs are required for infection to take place.

Polyprotein Processing

This strategy is adopted by viral genera such as the Potyviridae and Tymoviridae. The ribosome translates a single protein from the viral genome. Within the polyprotein is an enzyme (or enzymes) with proteinase function that is able to cleave the polyprotein into the various single proteins or just cleave away the protease, which can then cleave other polypeptides producing the mature proteins.

Well Understood Plant Viruses

Tobacco mosaic virus (TMV) and Cauliflower mosaic virus (CaMV) are frequently used in plant molecular biology. Of special interest is the CaMV 35S promoter, which is a very strong promoter most frequently used in plant transformations.

3

Organic Agriculture and Soil Biodiversity

Living Soils For Agriculture

Soils contain enormous numbers of diverse living organisms assembled in complex and varied communities. Soil biodiversity reflects the variability among living organisms in the soil - ranging from the myriad of invisible microbes, bacteria and fungi to the more familiar macro-fauna such as earthworms and termites.

Plant roots can also be considered as soil organisms in view of their symbiotic relationships and interactions with other soil components. These diverse organisms interact with one another and with the various plants and animals in the ecosystem, forming a complex web of biological activity.

Environmental factors, such as temperature, moisture and acidity, as well as anthropogenic actions, in particular, agricultural and forestry management practices, affect to different extents soil biological communities and their functions.

Soil organisms contribute a wide range of essential services to the sustainable functioning of all ecosystems. They act as the primary driving agents of nutrient cycling, regulating the dynamics of soil organic matter, soil carbon sequestration and greenhouse gas emissions; modifying soil physical structure and water regimes; enhancing the amount and efficiency of nutrient acquisition by the vegetation; and enhancing plant health. These services are not only critical to the functioning of natural ecosystems but constitute an important resource for sustainable agricultural systems.

Healthy Soils From Agriculture

Capturing the benefits of soil biological activity for agricultural production requires adhering to the following ecological principles:

- Supply organic matter. Each type of soil organism occupies a different niche in the web of life and favours a different substrate and nutrient source. Most soil organisms rely on organic matter for food; thus a rich supply and varied source of organic matter will generally support a wider variety of organisms.
- Increase plant varieties. Crops should be mixed and their spatial-temporal distribution varied, to create a greater diversity of niches and resources that stimulate soil biodiversity. For example diverse habitats support complex mixes of soil organisms, and through crop rotation or inter-cropping, it is possible to encourage the presence of a wider variety of organisms, improve nutrient cycling and natural processes of pest and disease control.
- Protect the habitat of soil organisms. The activity of soil biodiversity can be stimulated by improving soil living conditions, such as aeration, temperature, moisture, and nutrient quantity and quality. In this regard, reduced soil tillage and minimized compaction - and refraining chemical use - are of particular note.

Improvement in agricultural sustainability requires, alongside effective water and crop management, the optimal use and management of soil fertility and soil physical properties. Both rely on soil biological processes and soil biodiversity. This calls for the widespread adoption of management practices that enhance soil biological activity and thereby build up long-term soil productivity and health.

Adaptation and further development of soil biodiversity management into sustainable land management practices requires solutions that pay adequate consideration to the synergies between the soil ecosystem and its productive capacity and agro-ecosystem health. One practical example of holistic agricultural management systems that promote and enhance agro-ecosystem health, including biodiversity, biological cycles and soil biological activity is organic agriculture.

Organic Agriculture Nurtures Soil Biodiversity

Scientific research has demonstrated that organic agriculture significantly increases the density and species of soil's life. Suitable conditions for soil fauna and flora as well as soil forming and conditioning and nutrient cycling are encouraged by organic practices such as: manipulation of crop rotations and strip-cropping; green manuring and organic fertilization (animal manure, compost, crop residues); minimum tillage; and of course, avoidance of pesticides and herbicides use.

Benefits of organic management on soil biological activity are summarized below:

- Abundant arthropods and earthworms. Organic management increases the abundance and species richness of beneficial arthropods living above ground and earthworms, and thus improves the growth conditions of crops. More abundant predators help to control harmful organisms (pests). In organic systems the density and abundance of arthropods, as compared to conventional systems, has up to 100% more carabids, 60-70% more staphylinids and 70-120% more spiders. This difference is explained by prey deficiency due to pesticide influence as well as by a richer weed flora in the standing crop that is less dense than in conventional plots. In the presence of field margins and hedges, beneficial arthropods are further enhanced, as these habitats are essential for over-wintering and hibernation. The biomass of earthworms in organic systems is 30-40% higher than in conventional systems, their density even 50-80% higher. Compared to the mineral fertilizer system, this difference is even more pronounced.
- High occurrence of symbionts. Organic crops profit from root symbioses and are better able to exploit the soil. On average, mycorrhizal colonization of roots is highest in crops of unfertilized systems, followed by organic systems. Conventional crops have colonization levels that are 30% lower. The most intense mycorrhizal root colonization is found in grass-clover, followed by the vetch rye intercrop. Roots of winter wheat are scarcely colonized. Even when all soils are inoculated with active micorrhizae, colonization is enhanced in organic soil. This indicates that, even at an inoculum in surplus, soil nutrients at elevated levels and plant protection suppress symbiosis. This underlines the importance of appropriate living conditions for specific organisms.
- High occurrence of micro-organisms. Earthworms work hand in hand with fungi, bacteria, and numerous other microorganisms

in soil. In organically managed soils, the activity of these organisms is higher. Micro-organisms in organic soils not only mineralize more actively, but also contribute to the build up of stable soil organic matter (there is less untouched straw material in organic than in conventional soils). Thus, nutrients are recycled faster and soil structure is improved. The amount of microbial biomass and decomposition is connected: at high microbial biomass levels, little light fraction material remains undecomposed and vices versa.

- Microbial carbon. The total mass of micro-organisms in organic systems is 20-40% higher than in the conventional system with manure and 60-85% than in the conventional system without manure. The ratio of microbial carbon to total soil organic carbon is higher in organic system as compared to conventional systems. The difference is significant at 60 cm depth (at 80 cm depth, no difference is observed). Organic management promotes microbial carbon (and thus, soil carbon sequestration potential).
- Enzymes. Microbes have activities with important functions in the soil system: soil enzymes indicate these functions. The total activity of micro-organisms can be estimated by measuring the activity of a living cell-associated enzyme such as dehydrogenase. This enzyme plays a major role in the respiratory pathway. Proteases in soil, where most organic N is protein, cleave protein compounds. Phosphatases cleave organic phosphorus compounds and thus provide a link between the plant and the stock of organic phosphorus in the soil. Enzyme activity in organic soils is markedly higher than in conventional soils. Microbial biomass and enzyme activities are closely related to soil acidity and soil organic matter content.
- Wild flora. Large organic fields (over 15 ha) featured flora six times more abundant than conventional fields, including endangered varieties. In organic grassland, the average number of herb species was found to be 25 percent more than in conventional grassland, including some species in decline. Vegetation structure and plant communities in organic grassland are more even and more typical for a specific site than in conventionally managed systems. In particular, field margin strips of organic farms and semi-natural habitats conserve weed species listed as endangered or at risk of extinction. Animal grazing behaviour or routing activity (e.g.

pigs) was found important in enhancing plant species composition. Weeds (often sown in strips in organic orchards to reduce the incidence of aphids) influence the diversity and abundance of arthropods and flowering weeds are particularly beneficial to pollinators and parasitoids.

- High-energy efficiency. Organic agriculture follows the ecosystem theory of closed (or semi-closed) nutrient cycle on the farm. Organic land management allows the development of a relatively rich weed-flora as compared to conventional systems. Some "accompanying plants" of a crop are desired and considered useful in organic management. The presence of versatile flora attracts beneficial herbivores and other air-borne or above-ground organisms. Their presence improves the nourishment of predatory arthropods. When comparing diversity and the demand of energy for microbial maintenance (as indicated by the metabolic quotient), it becomes evident that diverse populations need less energy per unit biomass. A diverse microbial population, as present in the organic field plots, may divert a greater part of the available carbon to microbial growth rather than maintenance. In agricultural practice this may be interpreted as an increased turnover of organic matter with a faster mineralization and delivery of plant nutrients. Finally, more organic matter is diverted to build-up stable soil humus.
- Erosion control: Organic soil management improves soil structure by increasing soil activity and thus, reduces erosion risk. Organic matter has a positive effect on the development and stability of soil structure. Silty and loamy soils profit from organic matter by an enhanced aggregate structure. Organic matter is adsorbed to the charged surfaces of clay minerals. The negative charge decreases with increasing particle size. Silt is very susceptible to erosion since it is not charged, but organic matter layers on the silt surface favour aggregates with silt too.

Definition of Soil Microbiology & soil in view of Microbiology

Definition: It is branch of science/microbiology which deals with study of soil microorganisms and their activities in the soil.

Soil: It is the outer, loose material of earth's surface which is distinctly different from the underlying bedrock and the region which support plant life. Agriculturally, soil is the region which supports the plant life by providing mechanical support and nutrients required for growth. From the microbiologist view point, soil is one of the most dynamic sites of biological interactions in the nature. It is the region where most of the physical, biological and biochemical reactions related to decomposition of organic weathering of parent rock take place.

Components of Soil: Soil is an admixture *of* five major components viz. organic mater, mineral matter, soil-air, soil water and soil microorganisms/living organisms. The amount/ proposition of these components varies with locality and climate.

1. Mineral / Inorganic Matter: It is derived from parent rocks/bed rocks through decomposition, disintegration and weathering process. Different types of inorganic compounds containing various minerals are present in soil. Amongst them the dominant minerals are Silicon, Aluminum and iron and others like Carbon, Calcium Potassium, Manganese, Sodium, Sulphur, Phosphorus etc. are in trace amount. The proportion of mineral matter in soil is slightly less than half of the total volume of the soil.
2. Organic matter/components: Derived from organic residues of plants and animals added in the soil. Organic matter serves not only as a source of food for microorganisms but also supplies energy for the vital processes of metabolism which are characteristics of all living organisms. Organic matter in the soil is the potential source of N, P and S for plant growth. Microbial decomposition of organic matter releases the unavailable nutrients in available from. The proportion of organic matter in the soil ranges from 3-6% of the total volume of soil.
3. Soil Water: The amount of water present in soil varies considerably. Soil water comes from rain, snow, dew or irrigation. Soil water serves as a solvent and carrier of nutrients for the plant growth. The microorganisms inhabiting in the soil also require water for their metabolic activities. Soil water thus, indirectly affects plant growth through its effects on soil and microorganisms. Percentage of soil-water is 25% total volume of soil.
4. Soil air (Soil gases): A part of the soil volume which is not occupied by soil particles i.e. pore spaces are filled partly with soil water and partly with soil air. These two components (water & air) together only accounts for approximately half the soil's volume. Compared with atmospheric air, soil is lower in oxygen

and higher in carbon dioxide, because CO2 is continuous recycled by the microorganisms during the process of decomposition of organic matter. Soil air comes from external atmosphere and contains nitrogen, oxygen Co2 and water vapour (CO2 > oxygen). Co2 in soil air (0.3-1.0%) is more than atmospheric air (0.03%). Soil aeration plays important role in plant growth, microbial population, and microbial activities in the soil.

5. Soil microorganisms: Soil is an excellent culture media for the growth and development of various microorganisms. Soil is not an inert static material but a medium pulsating with life. Soil is now believed to be dynamic or living system.

Soil contains several distinct groups of microorganisms and amongst them bacteria, fungi, actinomycetes, algae, protozoa and viruses are the most important. But bacteria are more numerous than any other kinds of microorganisms. Microorganisms form a very small fraction of the soil mass and occupy a volume of less than one percent. In the upper layer of soil (top soil up to 10-30 cm depth i.e. Horizon A), the microbial population is very high which decreases with depth of soil. Each organisms or a group of organisms are responsible for a specific change / transformation in the soil. The final effect of various activities of microorganisms in the soil is to make the soil fit for the growth & development of higher plants.

Living organisms present in the soil are grouped into two categories as follows.

1. Soil flora (micro flora) e.g. Bacteria, fungi, Actinomycetes, Algae and
2. Soil fauna (micro fauna) animal like eg. Protozoa, Nematodes, earthworms, moles, ants, rodents.

Relative proportion / percentage of various soil microorganisms are: Bacteria-aerobic (70%), anaerobic (13 %), Actinomycetes (13%), Fungi /molds (03 %) and others (Algae Protozoa viruses) 0.2-0.8 %. Soil organisms play key role in the nutrient transformations.

Scope and Importance of Soil Microbiology

Living organisms both plant and animal types constitute an important component of soil. Though these organisms form only a fraction (less than one percent) of the total soil mass, but they play important role in supporting plant communities on the earth surface. While studying the scope and importance of soil microbiology, soil-plant-animal ecosystem as such must be taken into account. Therefore, the scope and importance of soil microbiology, can be understood in better way by studying aspects like:

1. Soil as a living system
2. Soil microbes and plant growth
3. Soil microorganisms and soil structure
4. Organic matter decomposition
5. Humus formation
6. Biogeochemical cycling of elements
7. Soil microorganisms as bio-control agents
8. Soil microbes and seed germination
9. Biological N2 fixation
10. Degradation of pesticides in soil.

Soil as a Living System

Soil inhabit diverse group of living organisms, both micro flora (fungi, bacteria, algae and actinomycetes) and micro-fauna (protozoa, nematodes, earthworms, moles, ants). The density of living organisms in soil is very high i.e. as much as billions / gm of soil, usually density of organisms is less in cultivated soil than uncultivated / virgin land and population decreases with soil acidity. Top soil, the surface layer contains greater number of microorganisms because it is well supplied with Oxygen and nutrients. Lower layer / subsoil is depleted with Oxygen and nutrients hence it contains fewer organisms. Soil ecosystem comprises of organisms which are both, autotrophs (Algae, BOA) and heterotrophs (fungi, bacteria). Autotrophs use inorganic carbon from CO2 and are “primary producers” of organic matter, whereas heterotrophs use organic carbon and are decomposers/ consumers.

Soil Microbes and Plant Growth

Microorganisms being minute and microscopic, they are universally present in soil, water and air. Besides supporting the growth of various biological systems, soil and soil microbes serve as a best medium for plant growth. Soil fauna & flora convert complex organic nutrients into simpler inorganic forms which are readily absorbed by the plant for growth. Further, they produce variety of substances like IAA, gibberellins, antibiotics etc. which directly or indirectly promote the plant growth.

Soil microbes and soil structure

Soil structure is dependent on stable aggregates of soil particles- Soil organisms play important role in soil aggregation. Constituents of soil are viz. organic matter, polysaccharides, lignins and gums, synthesized by soil microbes plays important role in cementing / binding of soil particles. Further, cells and mycelial strands of fungi and actinomycetes, Vormicasts from earthworm is also found to play important role in soil aggregation. Different soil microorganisms, having soil aggregation / soil binding properties are graded in the order as fungi > actinomycetes > gum producing bacteria > yeasts.

Examples are: Fungi like *Rhizopus, Mucor, Chaetomium, Fusarium, Cladasporium, Rhizoctonia, Aspergillus, Trichoderma* and Bacteria like *Azofobacler, Rhizobium Bacillus* and *Xanlhomonas.*

Soil Microbes and Organic Matter Decomposition

The organic matter serves not only as a source of food for microorganisms but also supplies energy for the vital processes of metabolism that are characteristics of living beings. Microorganisms such as fungi, actinomycetes, bacteria, protozoa etc. and macro organisms such as earthworms, termites, insects etc. plays important role in the process of decomposition of organic matter and release of plant nutrients in soil. Thus, organic matter added to the soil is converted by oxidative decomposition to simpler nutrients / substances for plant growth and the residue is transformed into humus. Organic matter / substances include cellulose, lignins and proteins (in cell wall of plants), glycogen (animal tissues), proteins and fats (plants, animals). Cellulose is degraded by bacteria, especially those of genus *Cytophaga* and other genera *(Bacillus, Pseudomonas, Cellulomonas, and Vibrio Achromobacter)* and fungal genera *(Aspergillus, Penicilliun, Trichoderma, Chactomium, Curvularia).* Lignins and proteins are partially digested by fungi, protozoa and nematodes. Proteins are degraded to individual amino acids mainly by fungi, *actinomycetes* and *Clostridium.* Under unaerobic conditions of waterlogged soils, methane are main carbon containing product which is produced by the bacterial genera (strict anaerobes) *Methanococcus, Methanobacterium* and *Methanosardna.*

Soil Microbes and Humus Formation

Humus is the organic residue in the soil resulting from decomposition of plant and animal residues in soil, or it is the highly complex organic residual matter in soil which is not readily degraded by microorganism, or it is the soft brown/dark coloured amorphous substance composed of residual organic matter along with dead microorganisms.

Soil microbes and cycling of elements: Life on earth is dependent on cycling of elements from their organic / elemental state to inorganic compounds, then to organic compounds and back to their elemental states. The biogeochemical process through which organic compounds are broken down to inorganic compounds or their constituent elements is known "Mineralization", or microbial conversion of complex organic compounds into simple inorganic compounds & their constituent elements is known as mineralization.

Soil microbes plays important role in the biochemical cycling of elements in the biosphere where the essential elements (C, P, S, N & Iron etc.) undergo chemical transformations. Through the process of mineralization organic carbon, nitrogen, phosphorus, Sulphur, Iron etc. are made available for reuse by plants.

Soil Microbes and Biological N2 Fixation

Conversion of atmospheric nitrogen in to ammonia and nitrate by microorganisms is known as biological nitrogen fixation.

Fixation of atmospheric nitrogen is essential because of the reasons:

1. Fixed nitrogen is lost through the process of nitrogen cycle through denitrification.
2. Demand for fixed nitrogen by the biosphere always exceeds its availability.
3. The amount of nitrogen fixed chemically and lightning process is very less (i.e. 0.5%) as compared to biologically fixed nitrogen
4. Nitrogenous fertilizers contribute only 25% of the total world requirement while biological nitrogen fixation contributes about 60% of the earth's fixed nitrogen
5. Manufacture of nitrogenous fertilizers by "Haber" process is costly and time consuming.

The numbers of soil microorganisms carry out the process of biological nitrogen fixation at normal atmospheric pressure (1 atmosphere) and temp (around 20 °C).

Two groups of microorganisms are involved in the process of BNF.

A. Non-symbiotic (free living) and B. Symbiotic (Associative)

Non-symbiotic (free living): Depending upon the presence or absence of oxygen, non symbiotic N2 fixation prokaryotic organisms may be aerobic heterotrophs (*Azotobacter, Pseudomonas, Achromobacter)* or aerobic autotrophs *(Nostoc, Anabena, Calothrix, BGA)* and anaerobic heterotrophs *(Clostridium, Kelbsiella. Desulfovibrio)* or anaerobic Autotrophs *(Chlorobium, Chromnatium, Rhodospirillum, Meihanobacterium etc)*

Symbiotic (Associative)

The organisms involved are *Rhizobium, Bratfyrhizobium* in legumes (aerobic): *Azospirillum* (grasses), Actinonycetes frantic (with *Casuarinas,* Alder).

Soil Microbes as Biocontrol Agents

Several ecofriendly bioformulations of microbial origin are used in agriculture for the effective management of plant diseases, insect pests, weeds etc. eg: *Trichoderma* sp and *Gleocladium* sp are used for biological control of seed and soil borne diseases. Fungal genera *Entomophthora, Beauveria, Metarrhizium* and protozoa *Maltesia grandis. Malameba locustiae* etc are used in the management of insect pests. Nuclear polyhydrosis virus (NPV) is used for the control of *Heliothis* / American boll worm. Bacteria like *Bacillus thuringiensis, Pseudomonas* are used in cotton against Angular leaf spot and boll worms.

Degradation of Pesticides in Soil by Microorganisms

Soil receives different toxic chemicals in various forms and causes adverse effects on beneficial soil micro flora / micro fauna, plants, animals and human beings. Various microbes present in soil act as the scavengers of these harmful chemicals in soil. The pesticides/ chemicals reaching the soil are acted upon by several physical, chemical and biological forces exerted by microbes in the soil and they are degraded into non-toxic substances and thereby minimize the damage caused by the pesticides to the ecosystem. For example, bacterial genera like *Pseudomonas, Clostridium, Bacillus, Thiobacillus, Achromobacter etc. and* fungal genera like *Trichoderma, Penicillium, Aspergillus, Rhizopus, and Fusarium* are playing important role in the degradation of the toxic chemicals / pesticides in soil.

Biodegradation of Hydrocarbons

Natural hydrocarbons in soil like waxes, paraffin's, oils etc are degraded by fungi, bacteria and actinomycetes. E.g. ethane *(C2* H6) a paraffin hydrocarbon is metabolized and degraded by *Mycobacteria, Nocardia, Streptomyces Pseudomonas, Flavobacterium* and several fungi.

Soil Humus

Humus is the organic residue in the soil resulting from decomposition of plant and animal residues in soil, or it is the highly complex organic residual matter in soil which is not readily degraded by microorganism, or it is the soft brown/dark coloured amorphous substance composed of residual organic matter along with dead microorganisms.

Composition of Humus

In most soil, percentage of humus ranges from 2-10 percent, whereas it is up to 90 percent in peat bog. On average humus is composed of Carbon (58 %), Nitrogen (3-6 %, Av.5%), acids - humic acid, fulvic acid, humin, apocrenic acid, and C: N ratio 10:1 to 12:1. During the course of their activities, the microorganisms synthesize number of compounds which plays important role in humus formation.

Functions/Role of Humus:

1. It improves physical condition of soil
2. Improve water holding capacity of soil
3. Serve as store house for essential plant nutrients
4. Plays important role in determining fertility level of soil
5. It tend to make soils more granular with better aggregation of soil particles
6. Prevent leaching losses of water soluble plant nutrients
7. Improve microbial/biological activity in soil and encourage better development of plant-root system in soil
8. Act as buffering agent i.e. prevent sudden change in soil PH/ soil reaction
9. Serve as source of energy and food for the development of soil organisms
10. It supplies both basic and acidic nutrients for the growth and development of higher plants
11. Improves aeration and drainage by making the soil more porous
12. History of Soil Microbiology (1600 - 1920)
13. There is enough evidence in the literature to believe that microorganisms were the earliest of the living things that existed on this planet. Man depends on crop plants for his existence and crop plants in turn depend on soil and soil microorganisms for their nutrition. Scientists form the

beginning studied the microorganisms from water, air, soil etc. and recognized the role of microorganisms in natural processes and realized the importance of soil microorganisms in growth and development of plants.

Thus, we see that microorganisms have been playing a significant role long before they were discovered by man. Today, soil is considered to be the main source of scavenging the organic wastes through microbial action and is also a rich store house for industrial micro flora of great economic importance.

Unlike soil science whose origin can be traced back to Roman & Aryan times, soil microbiology is emerged as a distinct branch of soil science during first half of the 19th century. Some of the notable contributions made by several scientists in field of soil microbiology are highlighted in the following paragraphs.

A. V. Leeuwenhock (1673) discovered and described microorganisms through his own made first simple microscope with magnification of 200 to 300 times. He observed minute, moving objects which he called "animalcules" (small animals) which are now known as protozoa, fungi and bacteria. He for the first time made the authentic drawings of microorganisms (protozoa, bacteria, fungi).

Robert Hook (1635-1703) developed a compound microscope with multiple lenses and described the fascinating world of the microbes.

J. B. Boussingault (1838) showed that leguminous plants can fix atmospheric nitrogen and increase nitrogen content in the soil.

14. J. Von Liebig (1856) showed that nitrates were formed in soil due to addition of nitrogenous fertilizers in soil.

S. N. Winogradsky discovered the autotrophic mode of life among bacteria and established the microbiological transformation of nitrogen and sulphur. Isolated for the first time nitrifying bacteria and demonstrated role of these bacteria in nitrification (1890), further he demonstrated that free-living *Clostridium pasteuriamum* could fix atmospheric nitrogen (1893). Therefore, he is considered as "Father of soil microbiology".

W. B. Leismaan (1858) and M. S. Woronin (1866) demonstrated that root nodules in legumes were formed by a specific group of bacteria.

Jodin (1862, France) gave the first experimental evidence of elemental nitrogen fixation by microorganisms.

15. Robert Koch (1882) developed gelatin plate/ streak plate technique for isolation of specific type of bacteria in soil, formulated Koch's postulates to establish causal relationship between host - pathogen and disease.

R. Warington (1878) showed that nitrification in soil was a microbial process.

B. Frank i) discovered (1880) an actinomycetes "Frankia" (Actinorhizal symbiosis) inducing root nodules in non-legumes tress of genera *Alnus sp* and *Casurina* growing in temperate forests, ii) coined (1885) the term " Mycorrhiza" to denote association of certain fungal symbionts with plant roots (Mycorrhiza-A symbiotic association between a fungus and roots of higher plants. Renamed the genus Bacillus as Rhizobium (1889).

H. Hellriegel and H. Wilfarth (1886) showed that the growth of non-legume plant was directly proportional to the amount of nitrogen supplied, whereas, in legumes there was no relationship between the quantity of nitrogen supplied and extent of plant growth. They also suggested that bacteria in the root nodules of legumes accumulate atmospheric nitrogen and made it available to plants. Showed that a mutually beneficial association exists between bacteria (*Rhizobia*) and legume root and legumes could utilize atmospheric nitrogen (1988).

M. W. Beijerinck (1888) isolated root nodule bacteria in pure culture from nodules in legumes and named them as *Bacillus radicola* Considered as father of "Microbial ecology". He was the first Director of the Delft School of microbiology (Netherland).

Beijerinck and Winogradsky (1890) developed the enrichment culture technique for isolation of soil organisms, proved independently that transformation of nitrogen in nature is largely due to the activities of various groups of soil microorganisms (1891). Therefore, they are considered as "Pioneer's in soil bacteriology".

S. N. Winogadsky (1891) demonstrated the role of bacteria in nitrification and further in fill 1983 demonstrated that free living *Clostridium pasteurianum* could fix atmospheric nitrogen.

Omeliansky (1902) Found the Anaerobic Degradation of Cellulose by Soil Bacteria.

J. G. Lipman and P. E. Brown (1903, USA) studied ammonification of organic nitrogenous substances by soil microorganisms and developed the Tumbler or Beaker for studying different types of transformation in soil.

Hiltner (Germany, 1904) coined the term "Rhizosphere" to denote that region of soil which is subjected to the influence of plant roots. Rhizosphere is the region where soil and plant roots make contact.

Russel and Hutchinson (1909, England), proved the importance of protozoa controlling/ maintaining bacterial population and their activity in soil.

Conn (1918) developed "Direct soil examination" technique for studying soil microorganisms.

Rayner (192I) and Melin (1927) carried out the intensive study on Mycorrhiza.

16. History of Soil Microbiology (1921 – 20th Century)
17. S. A. Waksman published the book "Principles of soil Microbiology" and thereby encouraged the research in soil microbiology (1927). Studied the role of soil as the source of antagonistic organisms with special reference to soil actinomycetes (1942) and discovered the antibiotic "Streptomycin" produced by *Streptomyces griseus*, a soil actinomycets (1944).

Rossi (1929) and Cholondy (1930) developed "Contact Slide / Buried slide" technique for studying soil micro flora.

Van Niel (1931USA) studied chemoautotrophic bacteria and bacterial photosynthesis.

Bortels (1936) demonstrated the importance of molybdenum in accelerating nitrogen fixation by nodulating legumes.

Garrett (1936) established the school in UK on "Soil fungi and ecological classification".

Kubo (1939, Japan) showed/proved-the role and importance of "leghaemoglobin" (Red pigment) present in root nodules of legumes in nitrogen fixation.

Ruinen (1956) Dutch microbiologist coined the term "Phyllosphere" to denote the region of leaf influenced by microorganisms.

18. Alien et al (1980) (suggested that VAM fungi stimulate plant growth by physiological effects other than by enhancement of nutrient uptake.

Jensen (1942) developed the method of studying nodulation on agar media in test tubes.

Barbara Mosse and J. W. Gerdemann (1944) reported occurrence of VAM (vesicular-arbuscular Mycorrhiza) fungi (*Glomus, Aculopora genera*) in the roots of agricultural crop plants which helps in the mobilization of phosphate.

Starkey (1945) studied role of bacteria (Bacillus and Clostridium) in the transformation of iron.

Barker (1945) studied anaerobic fermentation by methane bacteria (Methanococcus, Methanosarcina)

Thornton, (1947), studied root nodule bacteria form clovers.

Virtanen (1947) studied chemistry and mechanism of leghaemoglobin in nitrogen fixation.

Nutman (1948 England) studied hereditary mechanism of root nodulation in legumes.

Burris and Wilson (1957) developed the "Isotope technique" to quantify the amount of nitrogen fixed and further isolated and characterized the enzyme "Nitrogenase".

Bergersen (1957 Australia) elaborated the biochemistry of nitrogen fixation in legume root nodules.

Carnham (1960 USA) discovered nitrogen fixation by cell-free extract of *Clostridium pasteurianum*.

Alexander Fleming started the "School of soil microbiology" at Cornell University to study microbial aspects of pesticides degradation (1961) and developed the antibiotic "Penicillin" from the fungus *Penicillium notatum* (1929).

Date, Brockwell and Roughley (1962, Australia) developed the technique of bio-inoculants production & seed application.

Hardy & Associates (1968, USA) developed the technique of measurement of nitrogenase activity by acetylene-reduction test coupled with gas chromatography and thereby estimation of biological nitrogen fixation.

R J Swaby (1970, Australia) developed "Biosuper" containing rock phosphate sulphur and Thiobacillus which was used to enhance the phosphorus nutrition of plants.

Foog and Stewart (1970, UK) intensified the work on N2 fixing blue-green algae.

Trinick (1973, Australia) isolated Rhizobia from root nodule of genus Trema (Parasponia) which was an unique association of Rhizobium with non-leguminous plants causing root nodulation.

Dobereiner and associates (1975, Brazil) studied nitrogen fixing potential of Azospirillum in some tropical forage grasses like Digitaria, Panicum and some cereals like maize, sorghum, wheat, rye etc. in their roots. He reported four species of Azospirillum viz. *A. lipoferum, A. brasilense, A. amazonense* and *A. serpedica.* He coined the term "Associative Symbiosis" to denote the association between nitrogen fixing Azospirillum and cereal roots. Recently this terminology has been changed and renamed as "Diazotrophic Biocoenocis".

Challham and Associates (1978) isolated an actinomycetous endophyte *Frankia sp* from root nodules of *Camptonia peregrina* which is again an example of non-leguminous root nodulation.

Dommergues & associates (France and Senegal) had discovered / reported nodules on stem of *Sesbania rostrata* which could fix nitrogen and therefore this legume can be used as an excellent green manure crop in low land rice cultivation. Similarly they also discovered N2 fixing stem nodules on *Casurina sp* caused by Frankia, an actinomycete.

Louis Pasteur Proved the role of soil microorganisms in biochemical changes of elements. He also showed that decomposition of organic residues in soil was dependent on the nature of organic matter and environmental conditions.

Brefeld Introduced the practice of isolating soil fungi by "Single Cell" technique and cultivating / growing them on solid media. He used gelatin (first solidifying agent) in culture media as solidifying agent.

Gerretsen & Mulder (Holland) studied "Phosphate mobilization" by soil microorganisms and showed the importance of molybdenum in nitrogen metabolism by microorganisms.

Fritch, fogg & Stewart (UK) and lyengar (India) studied fixation by algae in general and micro algae in particular. They also intensified the work on N2 fixing BGA.

James Trappe and Don Marx worked on ectomycorrhiza, colonizing the roots of forest trees.

W. S. Cook, G. C. Papavizas, J. Baker and N.S. Kerr contributed to the field of biological control of plant pathogens using antagonistic organisms from soil. From the beginning of 20th century emphasis was given to the study of microorganisms in soil in relation to their physiology, ecology, interrelationship, role in soil processes and soil fertility. Further role of fungi and actinomycetes in cellulose decomposition was better understood and cellulose decomposing, sulphur oxidizing, iron bacteria etc were isolated from soil and studied in detail.

History of Soil Microbiology in India

During last few decades greater emphasis has been given on some of the important aspects in soil microbiology in India which are:

1. Characterization of N2 fixing *Azotobacter, Rhizobium, Beijerinckia,* BGA etc.
2. Studies on P- solubilizing bacteria and fungi, celluloytic microorganisms, silage production role of humic acid etc.
3. Establishment (1979) of All-India Coordinated Project (AICP) on BNF at IARI and field oriented work on BNF.
4. Standardization of methods of bio-inoculants application to seed and soil.
5. Seed bacterization and response of crops to bio-inoculants.

Some of the most important contributions made on the different aspects in the field of soil microbiology by the scientists and research institutes in the country are highlighted in the following paragraphs:

C. N. Acharya (1940) contributed towards the better utilization of Agricultural wastes for the production of biogas & compost.

Sundara Rao (1962) established the "Division of Microbiology" at IARI New Delhi.

Madhok (Punjab) introduced the practice of using bacterial cultures for berseem.

Sanyasi Raju & Rajagopalan (Coimbatore) initiated the research work on root nodulation in legumes at Madras, Agil. College.

P. K. Dey (West Bengal) worked on free living N2 fixing organisms viz *Azotobacter, Beijerinckia* and BGA in rice fields and discovered N2 fixation by BGA in paddy.

M.O.P. lyengar (Madras Univ.) laid foundation stone of algal research in India.

Sadasivan (Madras) and Saxena (Allahabad) studied ecology and physiology of soil fungi along with rhizosphere phenomenon.

Singh B.N. pioneering research on soil protozoa in India.

Bhar J. V. (Bangalore) initiated work on the role of earthworms in the maintenance of soil fertility, biological nitrogen fixation and microbiology of phyllosphere.

Thirumalacher (Hindustan Antibiotics, Pune) developed antifungal antibiotics like Haymycin and Aureofungin.

Nandi (Bose Res. Institute, Calcutta) worked on production technology of antibiotics and bacterial fertilizers (Biofertilizers).

Desikachray (Madras) studied taxonomy of BGA in India.

Thomas (BARC, Mumbai) studied physiology of algae in India.

Raja Rammohan Rao (CRRI, Cuttak) studied on rhizosphere nitrogen fixation phenomenon.

Bhagyaraj (GKVK, Bangalore) studied Mycorrhiza and N2 fixation interactions.

Verma (JNU, Delhi) studied / worked on sulphur metabolism.

Subramaniam & Mahadevan (Univ. Madras) studied fundamental aspects of N2 fixation.

Modi, Sushil Kumar, Das and Thomas carried research on "Genetics of "Nif" gene in relation to BNF by *Rhizobium, Azospirillum* and *Kelbsiella.*

Bharadwaj (Palampur) studied / worked on microbiology of organic matter decomposition & role of celluloytic microorganisms.

Gaur (IARI) and Mishra (Hissar) studied the role of celluloytic microorganisms in accelerating the process of composting and compost making.

Karla and Garcha (Ludhiana) studied the phenomenon of cellulose degradation and legume bacteriology.

Ranganathan & Nellakantan (NDRI, Karnal) worked on silage microbiology and process of anaerobic decomposition in biogas production.

Vadher, Gupta, Sethunatathan and Raghu studied role of soil enzymes and microbiology of pesticide degradation in soil.

Dart & Wani (non symbiotic N2 fixation), Thomas, Kumar Rao, Nambiar and Rupela (symbiotic N2 fixation) and Krishna (VAM fungi), these scientist at ICRISAT, Hyderabad work on symbiotic & non-symbiotic N2 fixation in gram, groundnut, arhar, sorghum and millets.

N. V. Joshi (1920) reported first isolation and identification *of Rhizobium* from different cultivated legumes

Gangulee and Madhok, studied physiology of *Rhizobium* and production of *Rhizobium* inoculants.

Sen and Pal (1957) studied solubilization of phosphate by soil microorganisms.

A. Sankaran (1958) standardized quality of legume inoculants for first time in India.

P. K. Dey and R. Bhattacharya isolated for the first time a new, non-symbiotic N2 fixing bacterium *Derixa gummosa* in the world.

V. Iswaran (1959) reported the use of Indian peat as carrier for Biofertilizers production.

Dube J. N. (1975) reported coal (wood-coal), an alternative to peat as carrier material for biofertilizer production.

Types of Microorganisms in Soil

Living organisms both plants and animals, constitute an important component of soil. The pioneering investigations of a number of early microbiologists showed for the first time that the soil was not an inert static material but a medium pulsating with life. The soil is now believed to be a dynamic or rather a living system, containing a dynamic population of organisms/microorganisms. Cultivated soil has relatively more population of microorganisms than the fallow land, and the soils rich in organic matter contain much more population than sandy and eroded soils. Microbes in the soil are important to us in maintaining soil fertility / productivity, cycling of nutrient elements in the biosphere and sources of industrial products such as enzymes, antibiotics, vitamins, hormones, organic acids etc. At the same time certain soil microbes are the causal agents of human and plant diseases.

The soil organisms are broadly classified in to two groups viz soil flora and soil fauna, the detailed classification of which is as follows.

Soil Organisms

A. Soil Flora

   a) Microflora: 1. Bacteria 2. Fungi, Molds, Yeast, Mushroom 3. Actinomycetes, Stretomyces 4. Algae eg. BGA, Yellow Green Algae, Golden Brown Algae.

      1. Bacteria is again classified in I) Heterotrophic eg. symbiotic & non - symbiotic N2 fixers, Ammonifier, Cellulose Decomposers, Denitrifiers II) Autrotrophic

eg. Nitrosomonas, Nitrobacter, Sulphur oxidizers, etc.

b) Macroflora: Roots of higher plants

B. Soil Fauna

a) Microfauna: Protozoa, Nematodes

b) Macrofauna: Earthworms. moles, ants & others.

As soil inhabit several diverse groups of microorganisms, but the most important amongst them are: bacteria, actinomycetes, fungi, algae and protozoa. The characteristics and their functions / role in the soil are described in the next topics.

Soil Microorganism: Bacteria

Amongst the different microorganisms inhabiting in the soil, bacteria are the most abundant and predominant organisms. These are primitive, prokaryotic, microscopic and unicellular microorganisms without chlorophyll. Morphologically, soil bacteria are divided into three groups viz *Cocci* (round/spherical), (rod-shaped) and *Spirilla I Spirllum* (cells with long wavy chains). *Bacilli* are most numerous followed by Cocci *and Spirilla in* soil.

The most common method used for isolation of soil bacteria is the "dilution plate count" method which allows the enumeration of only viable/living cells in the soil. The size of soil bacteria varies from 0.5 to 1.0 micron in diameter and 1.0 to 10.0 microns in length. They are motile with locomotory organs flagella.

Bacterial population is one-half of the total microbial biomass in the soil ranging from 1,00000 to several hundred millions per gram of soil, depending upon the physical, chemical and biological conditions of the soil.

Winogradsky (1925), on the basis of ecological characteristics classified soil microorganisms in general and bacteria in particular into two broad categories i.e. Autochnotus (Indigenous species) and the Zymogenous (fermentative). Autochnotus bacterial population is uniform and constant in soil, since their nutrition is derived from native soil organic matter (eg. *Arthrobacter* and *Nocardia* whereas Zymogenous bacterial population in soil is low, as they require an external source of energy, eg. *Pseudomonas & Bacillus*. The population of Zymogenous bacteria increases gradually when a specific substrate is added to the soil. To this category belong the cellulose decomposers, nitrogen utilizing bacteria and ammonifiers.

As per the system proposed in the Bergey's Manual of Systematic Bacteriology, most of the bacteria which are predominantly encountered in soil are taxonomically included in the three orders, Pseudomonadales, Eubacteriales and Actinomycetales of the class Schizomycetes. The most common soil bacteria belong to the genera *Pseudomonas, Arthrobacter, Clostridium Achromobacter, Sarcina, Enterobacter* etc. The another group of bacteria common in soils is the Myxobacteria belonging to the genera *Micrococcus, Chondrococcus, Archangium, Polyangium, Cyptophaga.*

Bacteria are also classified on the basis of physiological activity or mode of nutrition, especially the manner in which they obtain their carbon, nitrogen, energy and other nutrient requirements. They are broadly divided into two groups i.e. a) Autotrophs and b) Heterotrophs

1. Autotrophic bacteria are capable synthesizing their food from simple inorganic nutrients, while heterotrophic bacteria depend on pre-formed food for nutrition. All autotrophic bacteria utilize Co2 (from atmosphere) as carbon source and derive energy either from sunlight (photoautotrophs, eg. *Chromatrum. Chlorobium. Rhadopseudomonas* or from the oxidation of simple inorganic substances present in soil (chemoautotrophs eg. *Nitrobacter, Nitrosomonas, Thiaobacillus).*
2. Majority of soil bacteria are heterotrophic in nature and derive their carbon and energy from complex organic substances/ organic matter, decaying roots and plant residues. They obtain their nitrogen from nitrates and ammonia compounds (proteins) present in soil and other nutrients from soil or from the decomposing organic matter. Certain bacteria also require amino acids, B- Vitamins, and other growth promoting substances also.

Functions / Role of Bacteria:

Bacteria bring about a number of changes and biochemical transformations in the soil and thereby directly or indirectly help in the nutrition of higher plants growing in the soil. The important transformations and processes in which soil bacteria play vital role are: decomposition of cellulose and other carbohydrates, ammonification (proteins ammonia), nitrification (ammonia-nitrites-nitrates), denitrification (release of free elemental nitrogen), biological fixation of atmospheric nitrogen (symbiotic and non-symbiotic) oxidation and reduction of sulphur and iron compounds. All these processes play a significant role in plant nutrition,

Process/reaction	***Bacterial genera***
Cellulose decomposition (celluloytic bacteria) most cellulose decomposers are mesophilic	a. Aerobic : *Angiococcus, Cytophaga, Polyangium, Sporocytophyga, Bacillus, Achromobacter, Cellulomonas* b. *anaerobic: Clostridium Methanosarcina, Methanococcus*
Ammonification (Ammonifiers)	*Bacillus, Pseudomonas*
Nitrification (Nitrifying bacteria)	*Nitrosomonas, Nilrobacter Nitrosococcus*
Denitrification (Denitrifies)	*Achromobacter, Pseudomonas, Bacillus, Micrococcus*
Nitrogen fixing bacteria	a Symbiotic- Rhizobium, *Bradyrrhizobium* b Non-symbiotic: aerobic – Azotobacter *Beijerinckia* (acidic soils), anaerobic-*Clostridium*

Bacteria capable of degrading various plant residues in soil are:

Cellulose	***Hemicelluloses***	***Lignin***	***Pectin***	***Proteins***
Pseudomonas	Bacillus	Pseudomonas	Erwinia	Clostridium
Cytophaya	Vibrio	Micrococcus		Proteus
Spirillum	Pseudomonas	Flavobacteriumm		Pseudomonas
Actinomycetes	Erwinia	Xanthomonas		Bacillus
Cellulomonas		Streptomyces		

Soil Microorganism – Actinomycetes

These are the organisms with characteristics common to both bacteria and fungi but yet possessing distinctive features to delimit them into a distinct category. In the strict taxonomic sense, actinomycetes are clubbed with bacteria the same class of Schizomycetes and confined to the order Actinomycetales.

They are unicellular like bacteria, but produce a mycelium which is non-septate (coenocytic) and more slender, tike true bacteria they do not have distinct cell-wall and their cell wall is without chitin and cellulose (commonly found in the cell wall of fungi). On culture media unlike slimy distinct colonies of true bacteria which grow quickly, actinomycetes colonies grow slowly, show powdery consistency and stick firmly to agar surface. They produce hyphae and conidia / sporangia like fungi. Certain actinomycetes whose hyphae undergo segmentation resemble bacteria, both morphologically and physiologically.

Actinomycetes are numerous and widely distributed in soil and are next to bacteria in abundance. They are widely distributed in the soil, compost etc. Plate count estimates give values ranging from 10^4 to 10^8 per gram of soil. They are sensitive to acidity / low PH (optimum PH range 6.5 to 8.0) and waterlogged soil conditions. The population of actinomycetes increases with depth of soil even up to horizon ‘C’ of a soil profiler They are heterotrophic, aerobic and mesophilic (25-30 ^c) organisms and some species are commonly present in compost and manures are thermophilic growing at 55-65° c temperature (eg. Thermoatinomycetes, Streptomyces).

Actinomycetes belonging to the order of Actinomycetales are grouped under four families viz Mycobacteriaceae, Actinomycetaceae, Streptomycetaceae and Actinoplanaceae. Actinomycetous genera which are agriculturally and industrially important are present in only two families of Actinomycetaceae and Strepotmycetaceae.

In the order of abundance in soils, the common genera of actinomycetes are Streptomyces (nearly 70%), Nocardia and Micromonospora although Actinomycetes, Actinoplanes, Micromonospora and Streptosporangium are also generally encountered.

Functions / Role of actinomycetes:

1. Degrade/decompose all sorts of organic substances like cellulose, polysaccharides, protein fats, organic-acids etc.
2. Organic residues / substances added soil are first attacked by bacteria and fungi and later by actinomycetes, because they are slow in activity and growth than bacteria and fungi.
3. They decompose / degrade the more resistant and indecomposable organic substance/matter and produce a number of dark black to brown pigments which contribute to the dark colour of soil humus.
4. They are also responsible for subsequent further decomposition of humus (resistant material) in soil.
5. They are responsible for earthy / musty odor / smell of freshly ploughed soils.
6. Many genera species and strains (eg. Streptomyces if actinomycetes produce/synthesize number of antibiotics like Streptomycin, Terramycin, Aureomycin etc.
7. One of the species of actinomycetes *Streptomyces scabies* causes disease "Potato scab" in potato.

Soil Microorganism – Fungi

Fungi in soil are present as mycelial bits, rhizomorph or as different spores. Their number varies from a few thousand to a few -million per gram of soil. Soil fungi possess filamentous mycelium composed of individual hyphae. The fungal hyphae may be aseptate /coenocytic (Mastigomycotina and Zygomycotina) or septate (Ascomycotina, Basidiomycotina & Deuteromycotina).

As observed by C.K. Jackson (1975), most commonly encountered genera of fungi in soil are; *Alternaria, Aspergillus, Cladosporium, Cephalosporium Botrytis, Chaetomium, Fusarium, Mucor, Penicillium, Verticillium, Trichoderma, Rhizopus, Gliocladium, Monilia, Pythium,* etc. Most of these fungal genera belong to the sub-division Deuteromycotina / Fungi imperfeacta which lacks sexual mode of reproduction.

As these soil fungi are aerobic and heterotrophic, they require abundant supply of oxygen and organic matter in soil. Fungi are dominant in acid soils, because acidic environment is not conducive / suitable for the existence of either bacteria or actinomycetes. The optimum PH range for fungi lies-between 4.5 to 6.5. They are also present in neutral and alkaline soils and some can even tolerate PH beyond 9.0

Functions / Role of Fungi

1. Fungi plays significant role in soils and plant nutrition.
2. They plays important role in the degradation / decomposition of cellulose, hemi cellulose, starch, pectin, lignin in the organic matter added to the soil.
3. Lignin which is resistant to decomposition by bacteria is mainly decomposed by fungi.
4. They also serve as food for bacteria.
5. Certain fungi belonging to sub-division Zygomycotina and Deuteromycotina are predaceous in nature and attack on protozoa & nematodes in soil and thus, maintain biological equilibrium in soil.
6. They also plays important role in soil aggregation and in the formation of humus.
7. Some soil fungi are parasitic and causes number of plant diseases such as wilts, root rots, damping-off and seedling blights eg. *Pythium, Phyiophlhora, Fusarium, Verticillium* etc.
8. Number of soil fungi forms mycorrhizal association with the roots of higher plants (symbiotic association of a fungus with the roots of a higher plant) and helps in mobilization of soil phosphorus and nitrogen eg. Glomus, *Gigaspora, Aculospora,* (Endomycorrhiza) and *Amanita, Boletus, Entoloma, Lactarius* (Ectomycorrhiza).

Soil Microorganism – Algae

Algae are present in most of the soils where moisture and sunlight are available. Their number in soil usually ranges from 100 to 10,000 per gram of soil. They are photoautotrophic, aerobic organisms and obtain CO2 from atmosphere and energy from sunlight and synthesize their own food. They are unicellular, filamentous or colonial. Soil algae are divided in to four main classes or phyla as follows:

1. Cyanophyta (Blue-green algae)
2. Chlorophyta (Grass-green algae)
3. Xanthophyta (Yellow-green algae)
4. Bacillariophyta (diatoms or golden-brown algae)

Out of these four classes / phyla, blue-green algae and grass-green algae are more abundant in soil. The green-grass algae and diatoms are dominant in the soils of temperate region while blue-green algae predominate in tropical soils. Green-algae prefer acid soils while blue green algae are commonly found in neutral and alkaline soils. The most common genera of green algae found in soil are: *Chlorella, Chlamydomonas, Chlorococcum, Protosiphon* etc. and that of diatoms are *Navicula, Pinnularia. Synedra, Frangilaria.*

Blue green algae are unicellular, photoautotrophic prokaryotes containing Phycocyanin pigment in addition to chlorophyll. They do not posses flagella and do not reproduce sexually. They are common in neutral to alkaline soils. The dominant genera of BGA in soil are: *Chrococcus, Phormidium, Anabaena, Aphanocapra, Oscillatoria* etc. Some BGA posses specialized cells know as "Heterocyst" which is the sites of nitrogen fixation. BGA fixes nitrogen (non-symbiotically) in puddle paddy/water logged paddy fields (20-30 kg/ha/season). There are certain BGA which possess the character of symbiotic nitrogen fixation in association with other organisms like fungi, mosses, liverworts and aquatic ferns *Azolla, eg Anabaena-Azolla* association fix nitrogen symbiotically in rice fields.

Functions / role of algae or BGA:

1. Plays important role in the maintenance of soil fertility especially in tropical soils.
2. Add organic matter to soil when die and thus increase the amount of organic carbon in soil.
3. Most of soil algae (especially BGA) act as cementing agent in binding soil particles and thereby reduce/prevent soil erosion.
4. Mucilage secreted by the BGA is hygroscopic in nature and thus helps in increasing water retention capacity of soil for longer time/period.

5. Soil algae through the process of photosynthesis liberate large quantity of oxygen in the soil environment and thus facilitate the aeration in submerged soils or oxygenate the soil environment.
6. They help in checking the loss of nitrates through leaching and drainage especially in un-cropped soils.
7. They help in weathering of rocks and building up of soil structure.

Soil Microorganism – Protozoa

These are unicellular, eukaryotic, colourless, and animal like organisms (Animal kingdom). They are larger than bacteria and size varying from few microns to a few centimeters. Their population in arable soil ranges from 10,000 to 1,00,000 per gram of soil and are abundant in surface soil. They can withstand adverse soil conditions as they are characterized by "cyst stage" in their life cycle. Except few genera which reproduce sexually by fusion of cells, rest of them reproduces asexually by fission / binary fission. Most of the soil protozoa are motile by flagella or cilia or pseudopodia as locomotors organs. Depending upon the type of appendages provided for locomotion, protozoa are:

1. Rhizopoda (Sarcondia)
2. Mastigophora
3. Ciliophora (Ciliata)
4. Sporophora (not common Inhabitants of soil)

Class-Rhizopoda: Consists protozoa without appendages usually have naked protoplasm without cell-wall, pseudopodia as temporary locomotory organs are present some times. Important genera are *Amoeba, Biomyxa, Euglypha,* etc.

Class Mastigophora: Belongs flagellated protozoa, which are predominant in soil. Important genera are: *Allention, Bodo, Cercobodo, Cercomonas, Entosiphon Spiromonas, Spongomions* and *Testramitus.* Many members are saprophytic and some posses chlorophyll and are autotrophic in nature. In this respect, they resemble unicellular algae and hence are known as "Phytoflagellates".

The soil protozoa belonging to the class ciliate / ciliophora are characterized by the presence of cilia (short hair-like appendages) around their body, which helps in locomotion. The important soil inhabitants of this class are *Colpidium, Colpoda, Balantiophorus,*

Gastrostyla, Halteria, Uroleptus, Vortiicella, Pleurotricha etc. Protozoa are abundant in the upper layer (15 cm) of soil. Organic manures protozoa. Soil moisture, aeration, temperature and PH are the important factors affecting soil protozoa.

Function / Role of Protozoa

1. Most of protozoans derive their nutrition by feeding or ingesting soil bacteria belonging to the genera *Enterobacter, Agrobacterium, Bacillus, Escherichia, Micrococcus,* and *Pseudomonas* and thus, they play important role in maintaining microbial / bacterial equilibrium in the soil.
2. Some protozoa have been recently used as biological control agents against phytopathogens.
3. Species of the bacterial genera viz. *Enterobacter* and *Aerobacter* are commonly used as the food base for isolation and enumeration of soil protozoans.
4. Several soil protozoa cause diseases in human beings which are carried through water and other vectors, eg. Amoebic dysentery caused by *Entomobea histolytica.*

Factors Affecting Distribution, Activity and Population of Soil Microorganisms

Soil microorganisms (Flora & Fauna), just like higher plants depends entirely on soil for their nutrition, growth and activity. The major soil factors which influence the microbial population, distribution and their activity in the soil are

1. Soil fertility
2. Cultural practices
3. Soil moisture
4. Soil temperature
5. Soil aeration
6. Light
7. Soil PH (H-ion Concentration)
8. Organic matter
9. Food and energy supply
10. Nature of soil and
11. Microbial associations.

All these factors play a great role in determining not only the number and type of organism but also their activities. Variations in any one or more of these factors may lead to the changes in the activity of the organisms which ultimately affect the soil fertility level. Brief account of all these factors influencing soil micro flora / organisms and their activities is activities are discussed paragraphs.

Cultural Practices (Tillage): Cultural practices viz. cultivation, crop rotation, application of manures and fertilizers, liming and gypsum application, pesticide/fungicide and weedicide application have their effect on soil organism. Ploughing and tillage operations facilitate aeration in soil and exposure of soil to sunshine and thereby increase the biological activity of organisms, particularly of bacteria. Crop rotation with legume maintains the favourable microbial population balance, particularly of N2 fixing bacteria and thereby improve soil fertility.

Liming of acid soils increases activity of bacteria and actinomycetes and lowers the fungal population. Fertilizers and manures applied to the soil for increased crop production, supply food and nutrition not only to the crops but also to microorganisms in soil and thereby proliferate the activity of microbes.

Foliar or soil application of different chemicals (pesticides, fungicides, nematicides etc.) in agriculture are either degraded by the soil organisms or are liable to leave toxic residues in soil which are hazardous to cause profound reduction in the normal microbial activity in the soil.

Soil Fertility: Fertility level of the soil has a great influence on the microbial population and their activity in soil. The availability of N, P and K required for plants as well as microbes in soil determines the fertility level of soil. On the other hand soil micro flora has greater influence on the soil fertility level.

Soil Moisture: It is one of the important factors influencing the microbial population & their activity in soil. Water (soil moisture) is useful to the microorganisms in two ways i.e. it serve as source of nutrients and supplies hydrogen / oxygen to the organisms and it serve as solvent and carrier of other food nutrients to the microorganisms. Microbial activity & population proliferate best in the moisture range of 20% to 60%. Under excess moisture conditions / water logged conditions due to lack of soil aeration (Oxygen) anaerobic microflora become active and the aerobes get suppressed. While in the absence of adequate moisture in soil, some of microbes die out due to tissue dehydration and some of them change their forms into resting stages spores or cysts and tide over adverse conditions.

Therefore optimum soil moisture (range 20 to 60 %) must be there for better population and activity of microbes in soil.

Soil Temperature: Next to moisture, temperature is the most important environmental factor influencing the biological physical & chemical processes and of microbes, microbial activity and population in soil. Though microorganisms can tolerate extreme temperature (such as - 60 ° or + 60 u) conditions, but the optimum temperature range at which soil microorganisms can grow and function actively is rather narrow.

Depending upon the temperature range at which microorganisms can grow and function, are divided into three groups i.e. psychrophiles (growing at low temperature below 10 °C) Mesophiles (growing well in the temp range of 20 ° C to 45° C) and thermopiles (can tolerate temperature above 45° C and optimum 45-60°C).

Most of the soil microorganisms are mesophilic (25 to 40 °) and optimum temperature for most mesophiles is 37° C. True psychrophiles are almost absent in soil, and thermopiles though present in soil behaves like mesophiles. True thermopiles are more abundant in decaying manure and compost heaps where high temperature prevails.

Seasonal changes in soil temperature affect microbial population and their activity especially in temperate regions. In winter, when temperature is low (below 50° C), the number and activity of microorganisms falls down, and as the soils warms up in spring, they increases in number as well as activity. In general, population and activities of soil microorganisms are the highest in spring and lowest in winter season.

Soil air (Aeration): For the growth of microorganisms better aeration (oxygen and sometimes CO2) in the soil is essential. Microbes consume oxygen from soil air and gives out carbon dioxide. Activities of soil microbes is often measured in terms of the amount of oxygen absorbed or amount of Co2 evolved by the organisms in the soil environment. Under high soil moisture level / water logged conditions, gaseous exchange is hindered and the accumulation of Co4 occurs in soil air which is toxic to microbes. Depending upon oxygen requirements, soil microorganisms are grouped into categories viz aerobic (require oxygen for like processes), anaerobic (do not require oxygen) and microaerophilic (requiring low concentration / level of oxygen).

Light: Direct sunlight is highly injurious to most of the microorganisms except algae. Therefore upper portion of the surface soil a centimeter or less is usually sterile or devoid of microorganisms. Effect of sunlight is due to heating and increase in temperature (More than 45°)

Soil Reaction / Soil PH: Soil reaction has a definite influence / effect on quantitative and qualitative composite on of soil microbes. Most of the soil bacteria, blue-green algae, diatoms and protozoa prefer a neutral or slightly alkaline reaction between PH 4.5 and 8.0 and fungi grow in acidic reaction between PH 4.5 and 6.5 while actinomycetes prefer slightly alkaline soil reactions. Soil reactions also influence the type of the bacteria present in soil. For example nitrifying bacteria *(Nitrosomonas & Nitrobacter)* and diazotrophs like *Azotobacter* are absent totally or inactive in acid soils, while diazotrophs like *Beijerinckia, Derxia,* and sulphur oxidizing bacteria like *Thiobacillus thiooxidans* are active in acidic soils.

Soil Organic Matter: The organic matter in soil being the chief source of energy and food for most of the soil organisms, it has great influence on the microbial population. Organic matter influence directly or indirectly on the population and activity of soil microorganisms. It influences the structure and texture of soil and thereby activity of the microorganisms.

Food and energy supply: Almost all microorganisms obtain their food and energy from the plant residues or organic matter / substances added to the soil. Energy is required for the metabolic activities of microorganisms. The heterotrophs utilize the energy liberated during the oxidation of complex organic compounds in soil, while autotrophs meet their energy requirement form oxidation of simple inorganic compounds (chemoautotroph) or from solar radiation (Photoautotroph). Thus, the source of food and energy rich material is essential for the microbial activity in soil. The organic matter, therefore serves both as a source of food nutrients as well as energy required by the soil organisms.

Nature of Soil: The physical, chemical and physico-chemical nature of soil and its nutrient status influence the microbial population both quantitatively and qualitatively. The chemical nature of soil has considerable effect on microbial population in soil. The soils in good physical condition have better aeration and moisture content which is essential for optimum microbial activity. Similarly nutrients (macro and micro) and organic constituents of humus are responsible for absence or presence of certain type of microorganisms and their activity. For example activity and presence of nitrogen fixing bacteria is greatly influenced by the availability of molybdenum and absence of available phosphate restricts the growth of *Azotobacter*.

Microbial Associations/Interactions: Microorganisms interact with each other giving rise to antagonistic or symbiotic interactions. The association existing between one organism and another whether of symbiotic or antagonistic influences the population and activity of soil microbes to a great extent. The predatory habit of protozoa and some mycobacteria which feed on bacteria may suppress or eliminate certain bacteria. On the other hand, the activities of some of the microorganisms are beneficial to each other. For instance organic acids liberated by fungi, increase in oxygen by the activity of algae, change in soil reaction etc. favours the activity or bacteria and other organisms in soil.

Root Exudates: In the soil where plants are growing the root exudates also affects the distribution, density and activity of soil microorganism. Root exudates and sloughed off material of root surfaces provide an abundant source of energy and nutrients and thus directly or indirectly influence the quality as well as quantity of microorganisms in the rhizosphere region. Root exudates contain sugars, organic acids, amino acids, sterols, vitamins and other growth factors which have the profound effect on soil microbes.

Rhizosphere Concept and It's Historical Background

The root system of higher plants is associated not only with soil environment composed of inorganic and organic matter, but also with a vast community of metabolically active microorganisms. As living plants create a unique habitat around the roots, the microbial population on and around the roots is considerably higher than that of root free soil environment and the differences may be both quantitative and qualitative.

1. Rhizosphere: It is the zone/region of soil immediately surrounding the plant roots together with root surfaces, or it is the region where soil and plant roots make contact, or it is the soil region subjected to influence of plant roots and characterized by increased microbial.
2. Rhizoplane: Root surface along with the closely adhering soil particles is termed as rhizoplane.

Historical Background: Term "Rhizosphere" was introduced for the first time by the German scientist Hiltner (1904) to denote that region of soil which is subjected to the influence of plant roots.

The concept of "Rhizosphere Phenomenon" which shows the mutual interaction of roots and microorganisms was came into existence with the work of Starkey et al (1929), Clark (1939) and Rauath and Katznelson (1957).

N. V. Krassinikov (1934) found that free living nitrogen-fixing bacteria, *Azotobacter* were unable to grow in the wheat rhizosphere.

Starkey (1938) examined the rhizosphere region of some plant species and demonstrated the effect of root exudates on the predominance of bacterial population in particular and other soil microorganisms in general in the rhizosphere region. Thus, he put forth the concept of "Rhizosphere effect / phenomenon" for the first time.

F E Clark (1949) introduced / coined the term "Rhizoplane" to denote the root surface together with the closely adhering soil particles.

R. I. Perotti (1925) suggested the boundaries of the rhizosphere region and showed that it was bounded on one side by the general soil region (called as Edaphosphere) and on the other side by the root tissues (called Histosphere).

G. Graf and S. Poschenrieder (1930) divided the rhizosphere region into two general areas i.e. outer rhizosphere and inner rhizosphere for the purpose of describing the same site of microbial action.

H. Katznelson (1946) suggested the R:S ratio i.e. the ratio between the microbial population in the rhizosphere (R) and in the soil (S) to find out the degree or extent of plant roots effect on soil microorganisms. R: S ratio gives a good picture of the relative stimulation of the microorganisms in the rhizosphere of different plant species.

R: S ratio is defined as the ratio of microbial population per unit weight of rhizosphere soil (R), to the microbial population per unit weight of the adjacent non-rhizosphere soil (S)

A. G. Lochhead and H. Katznelson (1940) examined in detail the qualitative differences between the microflora of the rhizosphere and microflora of the non-rhizosphere region and reported that gram-negative, rod shaped and non-spore forming bacteria are abundant in the rhizosphere than in the non-rhizosphere soil

C. Thom and H. Humfeld (1932) found that corn roots in acidic soils yielded predominantly *Trichoderma* while roots from alkaline soils mainly contained *Penicillium.*

M J. Timonin (1940) reported some differences in the fungal types and population in the rhizosphere of cereals and legumes. R: S ratio of fungal population was believed to be narrow in most of the plant species, usually not exceeding 10.

E. A. Peterson and others (1958) reported that the plant age and soil type influence the nature of fungal flora in the rhizosphere, and the number of fungal population gradually increases with the age of plant.

M. Adati (1932) studied many crops and found that though actinomycetes were relatively less stimulated than bacteria, but in some cases the R: S ratio of actinomycetes was as high as 62.

R. Venkatesan and G. Rangaswami (1965) studied the rhizosphere effect in rice plant on bacteria, actinomycetes and fungi and reported that (i) for actinornycetes R: S was more (ranging from 0 to 25) depending on the age of plant roots and the dominant genera reported were Nocardia, (ii) R:S ratio reduced with the depth of soil.

E. A. Gonsalves and V. S. Yalavigi (1960) reported the presence of greater number of algae in the rhizosphere

J. W. Rouatt et al reported positive rhizosphere effect on protozoa, but a negative effect on algae in wheat plants.

Microorganisms in the Rhizosphere and Rhizosphere Effect

The rhizosphere region is a highly favourable habitat for the proliferation, activity and metabolism of numerous microorganisms. The rhizosphere microflora can be enumerated intensively by microscopic, cultural and biochemical techniques. Microscopic techniques reveal the types of organisms present and their physical association with the outer root tissue surface / root hairs. The cultural technique most commonly followed is “serial dilution and plate count method” which reveal the quantitative and qualitative population of microflora. At the same time, a cultural method shows the selective enhancement of certain categories of bacteria. The biochemical techniques used are designed to measure a specific change brought about by the plant or by the microflora. The rhizosphere effect on most commonly found microorganisms viz. bacteria, actinomycetes, fungi, algae and protozoa is being discussed herewith in the following paragraphs.

Bacteria: The greater rhizosphere effect is observed with bacteria (R: S values ranging from 10-20 or more) than with actinomycetes and fungi. Gram-negative, rod shaped, non-sporulating bacteria which respond to root exudates are predominant in the rhizosphere *(Pseudomonas, Agrobacterium)*. While Gram-positive, rods, Cocci and aerobic spore forming *(Bacillus, Clostridium)* are comparatively rare in the rhizosphere. The most common genera of bacteria are: *Pseudomonas, Arthrobacter, Agrobacterium, Alcaligenes, Azotobacter, Mycobacterium, Flavobacter, Cellulomonas, Micrococcus* and others have been reported to be either abundant or sparse in the rhizosphere. From the agronomic point of view, the abundance of nitrogen fixing and phosphate solubilizing bacteria in the rhizosphere assumes a great importance. The aerobic bacteria are relatively less in the rhizosphere because of the reduced oxygen levels due to root respiration. The bacterial population in the rhizosphere is enormous in the ranging form 10^8 to 10^9 per gram of rhizosphere soil. They cover about 4-10% of the total root area occurring profusely on the root hair region and rarely in the root tips. There is predominance of amino acids and growth factors required by bacteria, are readily provided by the root exudates in the region of rhizosphere.

Fungi: In contrast to their effects on bacteria, plant roots do not alter / enhance the total count of fungi in the rhizosphere. However, rhizosphere effect is selective and significant on specific fungal genera *(Fusarium, Verticillium, Aspergillus* and *Penicillium)* which are stimulated. The R:S ratio of fungal population is believed to be narrow in most of the plants, usually not exceeding to 10. The soil / serial dilution and plating technique used for the enumeration of rhizosphere fungi may often give erratic results as most of the spore formers produce abundant colonies in culture media giving a wrong picture / estimate *(eg Aspergilli* and *Penicillia).* In fact the mycelial forms are more dominant in the field. The zoospore / forming lower fungi such as *Phytophthora, Pythium, Aphanomyces* are strongly attracted to the roots in response to particular chemical compounds excreted by the roots and cause diseases under favourable conditions. Several fungi *eg Gibberella* and *fujikurio* produces *phytohormones* and influence the plant growth.

Actinomycetes, Protozoa and Algae: Stimulation of actinomycetes in the rhizosphere has not been studied in much detail so far. It is generally understood that the actinomycetes are less stimulated in the rhizosphere than bacteria. However, when antagonistic actinomycetes increase in number they suppress bacteria. Actinomycetes may also increase in number when antibacterial agents are sprayed on the crop. Among the actinomycete, the phosphate solublizers (eg. *Nocardia, Streptomyces)* have a dominant role to play.

As rule actinomycetes, protozoa and algae are not significantly influenced by their proximity to the plant roots and their R: S ratios rarely exceed 2 to 3: 1 and around roots of plants, R: S ratio for these microorganisms may go to high. Because of large bacterial community, an increase in the number or activity of protozoa is expected in the rhizosphere. Flagellates and amoebae are dominant and ciliates are rare in the region.

Classification of Bacteria or Prokaryotes – Volume I

The classification of bacteria or prokaryotes (of significance in agriculture and allied fields) given in "Bergey's Manual of Systematic Bacteriology" is listed as follows.

Section-I: The spirochetes, the two important families are i.e. i. Leptospiracear, important genus : Leptospira and ii. Spirochaetaceae, important genus is Spirochaets.

Section -2: Aerobic or microacerophilic, motile, helical or viborid gram negative bacteria. The important genera (of academic interest) in this section are: Azospirillium, Campylobacter and Badellovibrio. Azospirillium cells are vibrioid with single polar flagellum, present or found within the roots of grasses, wheat, corn and other plants or as free-living soil organisms. They are either aerobic or microerophilic and fix N2 within plant roots. Important species are A.limpoferum and A.brasilense.

Section-3: Non – motile, Gram negative curved bacteria: Family-Spirosomaceae, Genera: Spirosoma, Runella and Flectobacillus.

Section-4: Gram –negative, aerobic rods and cocci: It is on the largest section containing most diverse group of bacteria important in agriculture, such as:

Family: Pseudomonadaceae, Three Important genera are.........

Pseudomonas: Several species are pathogenic to humans, animals and plants; cause spoilage of meat and other foods, species like P. syringae, cause diseases like leaf spots, leaf stripe, wilt and necrosis, P. fluorescens, is common soil saprophyte that produce a fluorescent pigment.

Xanthomonas: Produce characteristics yellow pigment Xanthomonadin, all species are pathogenic to plants causing diseases such as spots, streaks, cankers, wilts and rots. Xanthomonads, produce exocellular polysaccharides i.e. xantham gums useful in industrial application.

Zoogloea: Cells are embedded in gelatinous matrix to from slimy masses with finger like morphology. Species are saprophytic, commonly found on tricking filter beds, in sewage treatment plants, where they oxidize the organic matter of the sewage.

Family: Azoobacteriacea

Two important genera are Azotobacter and Azomonas. The species are saprophytes found in soil, water and plant rhizosphere, fix N2 under aerobic conditions and forms desiccation resistant spores called "Cysts".

Family: Rhizobiaceae: Three important genera are Rhizobium, Brandyrhizobium and Agrobacterium. Bacteria of the genus Rhizobium and Brandyrhizobium fix atmospheric. N2 symbiotically in legumes by including root nodules. Whereas, species of the genus Agrobacterium, do not fix N2 but they are plant pathogenic inducing tumors in crown, roots and stems of dicotyledons.

Family: Methylococcaceae: Two important genera are Methyloccus and Methylomonas. Bacteria are obligate methane-Oxides, use methane gas as a sole carbon and energy source under aerobic and microacerophilic conditions.

Family: Acetobacteriaceae: Two important genera are Acetobacter and Glucanobacter Member of these two genera are saprophytes, found in sugar or alcohol, enriched acidic environments such as flower, fruits , bear , wine , vinegar, honey etc. They are industrially important. Acetobacter peritrichous are used to make vinegar and Glucanobacters (Polar Flagella) are involved in the manufacture of chemical like dihydroxyacetone, sorbose etc.

Section -5: Facultatively anaerobic Gram-negative rods-

Two important families and important genera in respective families are:

Family: Enterobacteriaceae, Important genera are............

Escherichia: E.g. E.coli, inhabitant of lower portion of the intestine of human and warm-blooded animals, caused gastrogenteritis and urinary tract infections.

Shigella: Species are pathogenic, causing bacillary dysenter in human called Shigellosis.

Salmonella: All species are pathogenic in humans causing enteric fevers typhoid and paratyphoid fevers gastroenteritis and septicaemia.

Enterobacter: Species occur in water, sewage, soil, meat plants and vegetables. Some species are opportunistic human pathogens.

Erwinia: Species mainly associated with plants, causing disease such as blights, cankers, die-back, leaf spot, wilts, discolouration of plant tissues and soft rots.

Yersinia: These are parasites of animals but can also cause infections in humans such as plague. (Y.pestis).

Family: Vibrionaceae

Important Genera are: Vobrio, Aeromonas

Most Vibrio species are harmless saprophytes, but some species are pathogenic in human. E.g. V. cholerae, causing cholera in humans.

Section-7: Dissimilatory Sulphate or Sulphur Reducing Bacteria: The bacteria or organism belonging to this section are obligate anaerobes using sulphate, sulphur or other oxidized sulphur compounds as election acceptors and reducing them to H2S. These are gram-negative found in mud and marine environments and in intestinal tract of humans and animals. The important genera are Desulfovibrio (Vibrioid and helical cells), Desulfococcus (Spherical cells) and Desulfosarcina.

Section-8: The Rickettsias and Chlamydias: These are tiny, non motile, Gram- negative bacteria. They are obligate parasites, able to grow only within host cells. The two important genera are Rickettisia and Coxiella. The species of Rickettisai caused diseases like Rocky Mountain spotted fever, classical typhus fever, scrub typhus and the single of species of genus Coxiella causes Q fever a type of pneumaonia (Coxiella burnetii).

Section-9: The Mycoplasma: These are very small organisms devoid of cell wall. Because of lack of cell wall, mycoplasmas are not inhibited by penicillin antibiotic; however they can be inhibited by antibiotics that affect protein synthesis (E.g. tetracycline or Chloramphenicol). They can be cultivated invitro (In laboratory) on synthetic media as facultative anaerobes or obligate anaerobes. Mycoplasmas are placed in the Division –Teericutes , class –Mollicutes and Order- Mycoplasmatales, containing three families, viz.

i) Mycoplasmataceae, genera Mycoplasma and Ureplasma,

ii) Acholeplasmatoceae, genus Acholeplasma

iii) Spiroplasmataceae, genera, Spiroplasma, Anaeroplasma and Thermoplasma

Species of the genus, Mycoplasma are pathogenic to human and animals, E.g. M. pneumoniae, causing primary atypical pneumonia in humans, members of the genus Ureplasma cause arthritis in humans, pneumonia and urogential disease in cattle. Citrus stubborn is one of the important disease caused by Spiroplasma (S. citri) in citrus.

Factors Affecting microbial Flora of the Rhizosphere / Rhizosphere Effect

The most important factors which affect / influence the microbial flora of the rhizosphere or rhizosphere effect are: soil type & its moisture, soil amendments, soil PH, proximity of root with soil, plant species, and age of plant and root exudates.

A. Soil type and its moisture: In general, microbial activity and population is high in the rhizosphere region of the plants grown in sandy soils and least in the high humus soils, and rhizosphere organisms are more when the soil moisture is low. Thus, the rhizosphere effect is more in the sandy soils with low moisture content.

B. Soil amendments and fertilizers: Crop residues, animal manure and chemical fertilizers applied to the soil cause no appreciable effect on the quantitative or qualitative differences in the microflora of rhizosphere. In general, the character of vegetation is more important than the fertility level of the soil.

C. Soil PH/ Rhizosphere PH: Respiration by the rhizosphere microflora may lead to the change in soil rhizosphere PH. If the activity and population of the rhizosphere microflora is more, then the PH of rhizcsphere region is lower than that of surrounding soil or non-rhizosphere soil. Rhizosphere effect for bacteria and protozoa is more in slightly alkaline soil and for that of fungi is more in acidic soils.

D. Proximity of root with Soil: Soil samples taken progressively closer to the root system have increasingly greater population of bacteria, and actinomycetes and decreases with the distance and depth from the root system. Rhizosphere effect decline sharply with increasing distance between plant root and soil.

E. Plant Species: Different plant species inhabit often some what variable microflora in the rhizosphere region. The qualitative and quantitative differences are attributed to variations in the rooting habits, tissue composition and excretion products. In general, legumes show / produce a more pronounced rhizosphere effect than grasses or cereals. Biennials, due to their long growth period exert more prolonged stimulation on rhizosphere effect than annuals.

F. Age of Plant: The age of plant also alter the rhizosphere microflora and the stage of plant maturity controls the magnitude of rhizosphere effect and degree of response to specific microorganisms. The rhizosphere microflora increases in number with the age of the plant and reaching at peak during flowering which is the most active period of plant growth and metabolism. Hence, the rhizosphere effect was found to be more at the time of flowering than in the seedling or full maturity stage of the plants. The fungal flora (especially, *Cellulolytic* and *Amylolytic)* of the rhizosphere usually increases even after fruiting and the onset of senescence due to accumulation of moribund tissue and sloughed off root parts / tissues: whereas, bacterial flora of the rhizosphere decreases after the flowering period and fruit setting.

G. Root / exudates /excretion: One of the most important factors responsible for rhizosphere effect is the availability of a great variety of organic substances at the root region by way of root exudates/excretions. The quantitative and qualitative differences in the microflora of the rhizosphere from that of general soil are mainly due to influences of root exudates. The spectrum of chemical composition root exudates varies widely, and hence their influence on the microflora also varies widely.

Root exudates are composed of the chemical substances like:

Sr. No	***Root Executes***	***Chemical Substances***
1	Amino Acids	All naturally occurring amino acids.
2	Organic acids	Acetic, butyric, citric, fumaric, lactic, malic, propionic, succinic etc.
3	Carbohydrates / sugars	Arabinose, fructose, galactose, glucose, maltose, mannose, oligosaccharides, raffinose, ribose, sucrose, xylose etc.
4	Nucleic acid derivatives	Adenine, cystidine, guanine, undine
5	Growth factors (phytohormones)	Biotin, choline, inositol, pyridoxine etc
6	Vitamins	Thiamine, nicotinic acid, biotin etc
7	Enzymes	Amylase, invertase, protease, phosphatase etc.
8	Other compounds	Auxins, glutamine, glycosides, hydrocyanic acid peptides, Uv-absorbing compounds, nematode attracting factors, spore germination stimulators, spore inhibitors etc.

The nature and amount of chemical substances thus exuded are dependent on the species of plant, plant age, inorganic nutrients, and temperature, light intercity, O2 / CO2 level, root injury etc. Another source of nutrients for the microorganisms in the rhizosphere region is the sloughed off root epidermis which exert selective stimulation effect on some specific groups of microorganisms. For instance, glucose and amino acids in the exudates readily attract Gram-negative rods which predominantly colonize the roots. Sugars and amino acids in the root exudates stimulate the germination of chlamydospores and other resting spores of fungi; stimulation effect of root exudates on plant pathogenic fungi, nematodes is also well known.

Alterations in Rhizosphere Microflora

Foliar application of various chemicals leads to alterations in the rhizosphere microflora by changing the pattern of root exudates. The pattern of the rhizosphere microflora i.e. numbers and species composition can be changed / altered by various factors, such as: (i) Soil amendments, (ii) Foliar application of fertilizers / nutrients, fungicides, insecticides and hormones and (iii) Bacterization / microbial seed inoculants.

Soil Amendments

Soil amendments with inorganic and organic fertilizers can alter the rhizosphere microflora and an understanding of the type of changes in the microflora can be useful in the indirect control of pathogens. Dwivedi and Chaube (1985) showed that amendment of soil with neem-cake can stimulate the activity of actinomycetes which results into the reduction of propagules of *Macrophomina phaseolina*. It is also known to control phytopathogenic nematodes in soil by stimulating nematode trapping fungi. Amendment of soil with castor and bean leaves stimulate the activity of *Trichoderma viride* and *Penicillium* in the rhizosphere leading to the control of *Sclerotium rolfsii*.

Foliar Application of Fertilizers and Agrochemicals

Translocation of photosynthete from leaves to roots takes place as a part of the normal metabolic activity in plants. Therefore, organic substances, including plant protection chemicals (fungicides, insecticides), growth regulators and plant nutrients applied to foliage / leaves get absorbed into the leaf tissue and further get translocated to roots along with photosynthates. Many workers have reported that foliar application with various chemicals cause marked alterations in the number and kind / qualities of microorganisms in the rhizosphere of several cereals and leguminous crop plants. Thus, such an approach can be used as a new tool in the biological control of root diseases, stimulation of activity of nitrogen-fixing bacteria and other beneficial microorganisms in the soil.

Seed Treatment with Bio Inoculants

Bio inoculants such as *Azotobacter, Beijerinckia, Azospirillum, Rhizobium or* P -solubilizing microorganisms (eg. *Bcillus, plymyxa, Azotobacter croococcum, Aspergillus niger, Penicillium digitatum* etc.) When applied to the seed / soil helps in the establishment of beneficial microorganisms in the rhizosphere region which will further benefit in plant growth, encourage inhibition of plant pathogenic organisms in the root vicinity and enrich the soil with added microbial bio-mass.

Associative and Antagonistic Activities in the Rhizosphere

In natural environments (eg. Soil, Air, Water etc.) a number of relationships exist between individual microbes, microbial species and between individual cells. The composition of microflora of any habitat (soil / rhizosphere) is governed by the biological equilibrium created by the associations and interactions of all individuals found in the community. In soil and rhizosphere region, many microorganisms live in close proximity and their interactions with each other may be associative or antagonistic.

Associative Interactions / Activities in Rhizosphere

The dependence of one microorganism upon another for extra-cellular products (eg. amino acids & growth promoting substances) can be regarded as an associative activity / effect in rhizosphere. There is an increase in the exudation of amino acids, organic acids and monosaccharide by plant roots in the presence of microorganisms. Gibberellins and gibberellin- like substances are known to be produced by bacterial genera viz *Azotobacter, Arthrobacter, Pseudomonas,* and *Agrobacterium* which are commonly found in the rhizosphere. Microorganisms also influence root hair development, mucilage secretion and lateral root development. Fungi inhabiting the root surface facilitate the absorption of nutrient by the roots.

Mycorrhiza is one of the best known associative / symbiotic interactions which exist between the roots of higher plants and fungi. This mycorrhizal association has been found to improve plant growth through better uptake of phosphorus and zinc from soil, suppression of root pathogenic fungi and nematodes. Another example is association between the bacterium *Rhizobium* and roots of legumes and *Azospirillum* with cereal crops (wheat, rye, bajara, maize etc).

Antagonistic Interactions / Activities in Rhizosphere

The biochemical qualities of root exudates and the presence of antagonistic microorganisms, plays important role in encouraging or inhibiting the soil borne plant pathogens in the rhizosphere region. Several mutualistic, communalistic, competitive and antagonistic interactions exist in the rhizosphere. The number and qualities of antagonistic microorganisms in the rhizosphere could be increased through artificial means such as fertilizer application, organic amendments, foliar spraying of chemicals etc.

Antagonistic microorganisms in the rhizosphere play an important role in controlling some of the soil borne plant pathogens. Stanier et al (1966) discovered the bacterial strain *Pseudomonas fluorescens* and the fluorescent pigments of this species in biological control of root pathogens. Strains of *P. fluorescence* are collectively called as "Fluorescent Pseudomonads". They produce variety of biologically active compounds such as plant growth substances, cyanides, antibiotics and iron chelating substances called "Siderophores" Rovira and Campbell (1975) , showed that bacterial strains of *P fluorescens* could lyse the hyphae of *Gaumannomyces graminis var. Tritici,* the causative agent of take-all disease of wheat. Fluorescent pseudomonads *(P. fluorescens, P. putida)* are known to produce iron chelating substances called Siderophores. These are low molecular weight, extra cellular, iron-binding agents produced by pseudomonads in response to low iron stress or when Fe3 is in short supply. Thus, iron stress triggers the formation of iron-binding ligands called siderophores. Siderophores contains the pigments Pyovirdin (Fluorescent) and Pyocyanin (non-Fluorescent) having iron chelating properties. Another pigment "Pseudobactin" is a fluorescent chelator of iron which is known to promote plant growth and inhibition of pathogenic bacteria in the rhizosphere. An antibiotic called "Pyrrolnitrin" reduces damping-off disease in cotton caused by *Rhizoctonia solani.* Several species of *Bacillus* are known to cause mycolysis in the rhizosphere. *eg. Fusarium oxysporum* hyphae are known to undergo lysis in soil due to these bacterial metabolites.

The successful antagonists among fungi are *Trichoderma* sp (T. viride and *T. harzianum, T. hamatum)* and *Gliocladium virens* which parasitize, lyse or kill the phytopathogenic fungi in the soil. Antifungal and antibacterial actinomycetes in the rhizosphere play an important role in controlling pathogenic fungi and bacteria, for example *Micromonospora globosa* is a potent antagonist of *Fusarium udum* causing wilt of pigeon pea. Amoebae are also known to play an antagonistic role in controlling soil fungi, eg. control of take-all disease of wheat caused by *Gaumannomyces graminis* through the use of Myxamoebae. There can also occur antagonisms between two fungi producing metabolite and interfering the growth of the other fungus as in case of *Peniophora* antagonizing *Heterobasidium*.

Rhizosphere in Relation to Plant Pathogens

Plant root exudates influence pathogenic fungi, bacteria and nematodes in various ways. The effect may be in the form of attraction of fungal zoospores, or bacterial cells towards the roots; stimulation of germination of dormant spores and hatching of cysts of nematodes. Root exudates may contain inhibitory substances preventing the establishment of pathogens. The balance between the rhizosphere microflora and plant pathogens and soil microflora and plant pathogens is important in host-pathogenic relationship. In this context, the biochemical qualities of root exudates and the presence of antagonistic micro-organisms plays an important role in the proliferation and survival of root infecting pathogens in soil either through soil fungi stasis, inhibition or antibiosis of pathogens in the rhizosphere.

Some of the most common interactions between plant roots and plant pathogenic microorganisms in the rhizosphere are discussed herewith.

A. Zoospore attraction: Amino acids, organic acids and sugars in the root exudates stimulate the movement and attraction of zoospores towards root of the plants. For example attraction of zoospores has been reported in *Phytophthora citrophthora* (Citrus roots), *P. parasitica* (tobacco roots) and *Pythium aphanidermatum* (pea root).

B. Spore germination: The spores or conidia of many pathogenic fungi such as *Rhizoctonia, Fusarium, Sclerotium, Pythium, Phytophthora* etc. have been stimulated to germinate by the root exudates of susceptible cultivars of the host plants. There are some reports on the selective stimulation of *Fusarium, Pseudomonas* and root infecting nematodes in the rhizosphere region of the respective susceptible hosts. This stimulus to germination is especially important to those plant pathogens which are not vigorous competitors and remain in resting stage due to shortage of nutrients or fungistasis. As a rule, germination and subsequent hyphal development are promoted by non host species and also by both susceptible and resistant cultivars of the host plants. The quantity and quality of microorganisms present in the rhizosphere of disease resistant crop varieties are significantly different from those of susceptible varieties.

C. Changes in morphology and physiology of host plant: Changes in the physiology and morphology of host plant influence the rhizosphere microflora through root exudations. Hence, significant changes in the rhizosphere microflora of diseased plants were reported which are attributed to the nature and severity of the disease. Systemic virus diseases cause marked changes in the plant morphology and physiology to drastically alter the rhizosphere microflora.

D. Increase in antagonists activity: Root exudates provide a food base for the growth of antagonistic organisms which plays an important role in controlling / suppressing some of the soil borne plant pathogens. Generally, rhizosphere of the resistant plant varieties harboure moer number of *Streptomyces* and *Trichoderma* than that of susceptible varieties. For example in the rhizosphere of pigeon pea varieties resistant to *Fusarium udum,* the population of *Streptomyces* was found more which inhibited the growth of the pathogen. High density of *Trichoderma viride* in the rhizosphere of Tomato varieties resistant to *Verticillium* wilt has been reported with its ability to reduce the severity of wilt in susceptible plants.

E. Inhibition of pathogen: Root exudates containing toxic substances such as glycosides and hydrocyanic acid may inhibit the growth ofpathogens in the rhizosphere. It has been reported that root exudates from resistant varieties of Flax (eg. Bison) excrete a glucoside which on hydrolysis produces hydrocyanic acid that inhibits Fusarium oxysporum, the flax root pathogen. Exudates of resistant pea reduce the germination of spores of Fusarium oxysporum.

In this light, the rhizosphere may be considered as a microbiological buffer zone in which the microflora serves to protect the plants against the attack of the pathogens.

F. Attraction of bacteria and nematodes: Root exudates attracts phytopathogenic bacteria and fungi in the rhizosphere for example *Agrobacterium tumefaciens* have been reported to be attracted to the roots of the host plants like peas, maize, onion, tobacco, tomato and cucumber.

Host root exudates also influence phytopathogenic nematodes in two ways: (i) though stimulation of egg-hatching process and (ii) attraction of larvae towards plant roots.

Soil Microorganisms in Cycling of Elements or Plant Nutrient

Soil microorganisms are the most important agents in the cycling / transformation of various elements (N, P, K, S, Iron etc.) in the biosphere; where the essential elements undergo cyclic alterations between the inorganic state as free elements in nature and the combined state in living organisms. Life on earth is dependent on the cycling of nutrient elements from their elemental states to inorganic compounds to organic compounds and back into their elemental states.

The microbes through the process of biochemical reactions convert / breakdown complex organic compounds into simple inorganic compounds and finally into their constituent elements. This process is known as "Mineralization".

Mineralization of organic carbon, nitrogen, phosphorus, sulphur and iron by soil microorganisms makes these elements available for reuse by plants. In the following paragraphs the cycling / transformations of some of the important elements are discussed.

The four most important cycles are mention below

A. Nitrogen Cycle
B. Sulphur Cycle / Sulphur Transformation
C. Phosphorus Cycle / Transformation
D. Iron Cycle / Transformation

Nitrogen Cycle

Although molecular nitrogen (N2) is abundant (i.e 78-80 % by volume) in the earth's atmosphere, but it is chemically inert and therefore, can not be utilized by most living organisms and plants. Plants, animals and most microorganisms, depend - on a source of combined or fixed nitrogen (eg. ammonia, nitrate) or organic nitrogen compounds for their nutrition and growth. Plants require fixed nitrogen (ammonia, nitrate) provided by microorganisms, but about 95 to 98 % soil nitrogen is in organic form (unavailable) which restrict the development of living organisms including plants and microorganisms. Therefore, cycling/transformation of nitrogen and nitrogenous compounds mediated by soil microorganisms is of paramount importance in supplying required forms of nitrogen to the plants and various nutritional classes of organisms in the biosphere. In nature, nitrogen exists in three different forms viz. gaseous / gas (78 to 80 % in atmosphere), organic (proteins and amino acids, chitins, nucleic acids and amino sugars) and inorganic (ammonia and nitrates).

Biological N2 Fixation:

A. Symbiotic: Eg. Rhizobium (Eubacteria) legumes, Frankia (Actinomycete) and Anabaena (cyanobacteria) non - legumes

B. Non Symbiotic:

1. Free Living: eg. Azobacter, Derxia, Bejerinkia, Rhodospirillum and BGA.
2. Associative: eg. Azospirillum, Acetobacter, Herbaspirillim.

Nutritional categories of N2 fixing Bacteria:

A. Heterotrops

B. Photoautotrophs

Nitrogen cycle is the sequence of biochemical changes form free atmospheric N2 to complex organic compounds in plant and animal tissues and further to simple inorganic compounds (ammonia, nitrate) and eventual release of molecular nitrogen (N2) back to the atmosphere is called "nitrogen cycle".

In this cycle a part of atmospheric nitrogen (N2) is converted into ammonia and then to amino acids (by soil microorganisms and plant-microbe associations) which are used for the biosynthesis of complex nitrogen-containing organic compound such as proteins, nucleic acids, amino sugars etc. The proteins are then degraded to simpler organic compounds viz. peptones and peptides into amino acids which are further degraded to inorganic nitrogen compounds like ammonia, nitrites and nitrates. The nitrate form of nitrogen is mostly used by plants or may be lost through leaching or reduced to gaseous nitrogen and subsequently goes into the atmosphere, thus completing the nitrogen cycle. Thus, the process of mineralization (conversion of organic form of nutrients to its mineral /inorganic form) and immobilization (process of conversion of mineral / inorganic form of nutrient elements into organic form) are continuously and simultaneously going on in the soil.

Several biochemical steps involved in the nitrogen cycle are:

1. Proteolysis
2. Ammonification
3. Nitrification
4. Nitrate reduction and
5. Denitrification.

Nitrogen Cycle: Proteolysis & Ammonification

Several biochemical steps involved in the nitrogen cycle are:

1. Proteolysis

2. Ammonification
3. Nitrification
4. Nitrate reduction and
5. Denitrification.

Proteolysis: Plants use the ammonia produced by symbiotic and non-symbiotic Nitrogen fixation to make their amino acids & eventually plant proteins. Animals eat the plants and convert plant proteins into animal proteins. Upon death, plant and animals undergo microbial decay in the soil and the nitrogen contained in their proteins is released. Thus, the process of enzymatic breakdown of proteins by the microorganisms with the help of proteolysis enzymes is known as "proteolysis".

The breakdown of proteins is completed in two stages. In first stage proteins are converted into peptides or polypeptides by enzyme "proteinases" and in the second stage polypeptides / peptides are further broken down into amino acids by the enzyme "peptidases".

Proteins ——————> Peptides ——————> Amino Acids

Proteinases Peptidases

The amino acids produced may be utilized by other microorganisms for the synthesis of cellular components, absorbed by the plants through mycorrhiza or may be de animated to yield ammonia.

The most active microorganisms responsible for elaborating the proteolytic enzymes (Proteinases and Peptidases) are *Pseudomonas, Bacillus, Proteus, Clostridium Histolyticum, Micrococcus, Alternaria, Penicillium* etc.

Ammonification (Ammo acid Degradation): Amino acids released during proteolysis undergo deamination in which nitrogen containing amino (-NH2) group is removed. Thus, process of deamination which leads to the production of ammonia is termed as "ammonification". The process of ammonification is mediated by several soil microorganisms. Ammonification usually occurs under aerobic conditions (known as oxidative deamination) with the liberation of ammonia (NH3) or ammonium ions (NH4) which are either released to the atmosphere or utilized by plants (paddy) and microorganisms or still under favourable soil conditions oxidized to form nitrites and then to nitrates.

The processes of ammonification are commonly brought about by *Clostridium* sp, *Micrococcus* sp, *Proteus sp.* etc. and it is represented as follows.

Alanine

CH3 CHNH2 COOH + 1/2 O2 ——> C H3COCOOH + NH3

Alanine deaminase Pyruvic acid ammonia

Nitrogen Cycle: Nitrification & Nitrate Reduction

Several biochemical steps involved in the nitrogen cycle are:

1. Proteolysis
2. Ammonification
3. Nitrification
4. Nitrate reduction and
5. Denitrification.

Nitrification: Ammonical nitrogen / ammonia released during ammonification are oxidized to nitrates and the process is called "nitrification". Soil conditions such as well aerated soils rich in calcium carbonate, a temperature below 30 ° C, neutral PH and less organic matter are favourable for nitrification in soil.

Nitrification is a two stage process and each stage is performed by a different group of bacteria as follows.

Stage I: Oxidation of ammonia of nitrite is brought about by ammonia oxidizing bacteria viz. *Nitrosomnonas europaea, Nitrosococcus nitrosus, Nitrosospira briensis, Nitrosovibrio* and *Nitrocystis* and the process is known as nitrosification. The reaction is presented as follows.

2 NH3 + 1/2O2 ————————> NO2 + 2 H + H2 O

Ammonia Nitrite

Stage II: In the second step nitrite is oxidized to nitrate by nitrite-oxidizing bacteria such as *Nitrobacter winogradsky .Nitrospira gracilis, Nirosococcus mobiiis* etc, and several fungi *(eg. Penicillium, Aspergillus) and* actinomycetes *(eg. Streptomyces, Nocardia).*

NO2 (-) + ½ O2 ————————> NO3

Nitrite ions Nitrate ions

The nitrate thus, formed may be utilized by the microorganisms, assimilated by plants, reduced to nitrite and ammonia or nitrogen gas or lost through leaching depending on soil conditions. The nitrifying bacteria (ammonia oxidizer and nitrite oxidizer) are aerobic gram-negative and chemoautotrophic and are the common inhabitants of soil, sewage and aquatic environment.

Nitrate Reduction: Several heterotrophic bacteria *(E. coli, Azospirillum)* are capable of converting nitrates to nitrites and nitrites to ammonia. Thus, the process of nitrification is reversed completely which is known as nitrate reduction. Nitrate reduction normally occurs under anaerobic soil conditions (water logged soils) and the overall process is as follows:

$$\underset{\text{Nitrate}}{HNO3 + 4\ H2} \xrightarrow[\text{Reductase}]{\text{Nitrate}} \underset{\text{ammonium}}{NH4} + 3\ H20$$

Nitrate reduction leading to production of ammonia is called "dissimilatory nitrate reduction" as some of the microorganisms assimilate ammonium for synthesis of proteins and amino acid.

Nitrogen Cycle: Denitrification

Several biochemical steps involved in the nitrogen cycle are:

1. Proteolysis
2. Ammonification
3. Nitrification
4. Nitrate reduction and
5. Denitrification.

Denitrification: This is the reverse process of nitrification. During denitrification nitrates are reduced to nitrites and then to nitrogen gas and ammonia. Thus, reduction of nitrates to gaseous nitrogen by microorganisms in a series of biochemical reactions is called "denitrification". The process is wasteful as available nitrogen in soil is lost to atmosphere. The overall process of denitrification is as follows:

$$\text{Nitrate} \xrightarrow{\text{NaR}} \text{Nitrite} \xrightarrow{\text{NiR}} \text{Nitric Oxide} \xrightarrow{\text{NoR}} \text{Nitrous Oxide} \xrightarrow{\text{NoR}} \text{Nitrogen gas}$$

This process also called dissimilatory nitrate reduction as nitrate nitrogen is completely lost into atmospheric air. In the soils with high organic matter and anaerobic soil conditions (waterlogged or ill-drained) rate of denitrification is more. Thus, rice / paddy fields are more prone to denitrification.

The most important denitrifying bacteria are *Thiobacillus denitrificans, Micrococcus denitrificans,* and species of *Pseudomonas, Bacillus, Achromobacter, Serrtatia paracoccus* etc.

Denitrification leads to the loss of nitrogen (nitrate nitrogen) from the soil which results into the depletion of an essential nutrient for plant growth and therefore, it is an undesirable process / reaction from the soil fertility and agricultural productivity. Although, denitrification is an undesirable reaction from agricultural productivity, but it is of major ecological importance since, without denitrification the supply of nitrogen including N2 of the atmosphere, would have not got depleted and No3 (which are toxic) would have accumulated in the soil and water.

Sulphur Cycle / Sulphur Transformations

Sulphur is the most abundant and widely distributed element in the nature and found both in free as well as combined states. Sulphur, like nitrogen is an essential element for all living systems. In the soil, sulphur is in the organic form (sulphur containing amino acids-cystine, methionine, proteins, polypeptides, biotin, thiamine etc) which is metabolized by soil microorganisms to make it available in an inorganic form (sulphur, sulphates, sulphite, thiosulphale, etc) for plant nutrition. Of the total sulphur present is soil only 10-15% is in the inorganic form (sulphate) and about 75-90 % is in organic form.

Cycling of sulphur is similar to that of nitrogen. Transformation / cycling of sulphur between organic and elemental states and between oxidized and reduced states is brought about by various microorganisms, specially bacteria- Thus "the conversion of organically bound sulphur to the inorganic state by microorganisms is termed as mineralization of sulphur". The sulphur / sulphate, thus released are either absorbed by the plants or escapes to the atmosphere in the form of oxides.

Various transformations of the sulphur in soil results mainly due to microbial activity, although some chemical transformations are also possible (eg. oxidation of iron sulphide) the major types of transformations involved in the cycling of sulphur are:

1. Mineralization
2. Immobilization
3. Oxidation and
4. Reduction

   1. Mineralization: The breakdown / decomposition of large organic sulphur compounds to smaller units and their conversion into inorganic compounds (sulphates) by the microorganisms. The rate of sulphur mineralization is about 1.0 to 10.0 percent / year.
   2. Immobilization: Microbial conversion of inorganic sulphur compounds to organic sulphur compounds.

3. Oxidation: Oxidation of elemental sulphur and inorganic sulphur compounds (such as h2S, sulphite and thiosulphale) to sulphate (SO4) is brought about by chemoautotrophic and photosynthetic bacteria.

When plant and animal proteins are degraded, the sulphur is released from the amino acids and accumulates in the soil which is then oxidized to sulphates in the presence of oxygen and under anaerobic condition (water logged soils) organic sulphur is decomposed to produce hydrogen sulphide (H2S). H2S can also accumulate during the reduction of sulphates under anaerobic conditions which can be further oxidized to sulphates under aerobic conditions,

$$\text{a) } 2S + 3O_2 + 2H_2O \longrightarrow 2H_2SO_4 \xrightarrow{\text{Ionization}} 2H^{+} + SO_4 \text{ (Aerobic)}$$

$$\text{b) } CO_2 + 2H_2S \xrightarrow{\text{Light}} (CH_2O) + H_2O + 2S$$

$$\text{OR } H_2 + S + 2CO_2 + H_2O \xrightarrow{\text{Light}} H_2SO_4 + 2(CH_2O) \text{ (anaerobic)}$$

The members of genus *Thiobacillus* (obligate chemolithotrophic, non photosynthetic) eg, *T. ferrooxidans* and *T. thiooxidans* are the main organisms involved in the oxidation of elemental sulphur to sulphates. These are aerobic, non-filamentous, chemosynthetic autotrophs. Other than *Thiobacillus,* heterotrophic bacteria *(Bacillus, Pseudomonas, and Arthrobacter)* and fungi *(Aspergillus, Penicillium)* and some actinomycetes are also reported to oxidize sulphur compounds. Green and purple bacteria (Photolithotrophs) of genera *Chlorbium, Chromatium. Rhodopseudomonas* are also reported to oxidize sulphur in aquatic environment.

Sulphuric acid produced during oxidation of sulphur and H: S is of great significance in reducing the PH of alkaline soils and in controlling potato scab and rot diseases caused by *Streptomyces* bacteria. The formation of sulphate / Sulphuric acid is beneficial in agriculture in different ways : i) as it is the anion of strong mineral acid (H2 SO4) can render alkali soils fit for cultivation by correcting soil PH. ii) solubilize inorganic salts containing plant nutrients and thereby increase the level of soluble phosphate, potassium, calcium, magnesium etc. for plant nutrition.

Reduction of Sulphate: Sulphate in the soil is assimilated by plants and microorganisms and incorporated into proteins. This is known as "assimilatory sulphate reduction". Sulphate can be reduced to hydrogen sulphide (H_2S) by sulphate reducing bacteria (eg. *Desulfovibrio* and *Desulfatomaculum)* and may diminish the availability of sulphur for plant nutrition. This is "dissimilatory sulphate reduction" which is not at all desirable from soil fertility and agricultural productivity view point.

Dissimilatory sulphate-reduction is favoured by the alkaline and anaerobic conditions of soil and sulphates are reduced to hydrogen sulphide. For example, calcium sulphate is attacked under anaerobic condition by the members of the genus *Desulfovibrio* and *Desulfatomaculum* to release H_2S.

$$CaSO_4 + 4H_2 \longrightarrow Ca(OH)_2 + H_2S + H_2O.$$

Hydrogen sulphide produced by the reduction of sulphate and sulphur containing amino acids decomposition is further oxidized by some species of green and purple phototrophic bacteria *(eg. Chlorobium, Chromatium)* to release elemental sulphur.

$$\underset{}{CO_2 + 2H_2 + H_2S} \xrightarrow[\text{Enzyme}]{\text{Light}} \underset{\text{Carbohydrate}}{(CH_2O)} + H_2O + \underset{\text{Sulphur}}{2\,S.}$$

The predominant sulphate-reducing bacterial genera in soil are *Desulfovibrio, Desulfatomaculum* and *Desulfomonas.* (All obligate anaerobes). Amongst these species *Desulfovibrio desulphuricans* are most ubiquitous, non-spore forming, obligate anaerobes that reduce sulphates at rapid rate in waterlogged / flooded soils. While species of *Desulfatomaculum* are spore forming, thermophilic obligate anaerobes that reduce sulphates in dry land soils. All sulphate-reducing bacteria excrete an enzyme called "desulphurases" or "bisulphate Reductase". Rate of sulphate reduction in nature is enhanced by increasing water levels (flooding), high organic matter content and increased temperature.

Phosphorus Cycle or Transformation

Phosphorus is only second to nitrogen as a mineral nutrient required for plants, animals and microorganisms. It is a major constituent of nucleic acids in all living systems essential in the accumulation and release of energy during cellular metabolism. This element is added to the soil in the form of chemical fertilizers, or in the form of organic phosphates present in plant and animal residues. In cultivated soils it is present in abundance (i.e. 1100 kg/ha), but most of which is not available to plants, only 15 % of total soil phosphorus is in available form. Both inorganic and organic phosphates exist in soil and occupy a critical position both in plant growth and in the biology of soil.

Microorganisms are known to bring a number of transformations of phosphorus, these include:

(i) Altering the solubility of inorganic compounds of phosphorus,

(ii) Mineralization of organic phosphate compounds into inorganic phosphates,

(iii) Conversion of inorganic, available anion into cell components i.e. an immobilization process and

(iv) Oxidation or reduction of inorganic phosphorus compounds. Of these mineralization and immobilization are the most important reactions / processes in phosphorus cycle.

Insoluble inorganic compounds of phosphorus are unavailable to plants, but many microorganisms can bring the phosphate into solution. Soil phosphates are rendered available either by plant roots or by soil microorganisms through secretion of organic acids (eg. lactic, acetic, formic, fumaric, succinic acids etc). Thus, phosphate-dissolving / solubilizing soil microorganisms (eg. species of *Pseudomonas, Bacillus, Micrococcus, Mycobacterium, Flavobacterium, Penicillium, Aspergillus, Fusarium* etc.) plays important role in correcting phosphorus deficiency of crop plants. They may also release soluble inorganic phosphate (H2PO4), into soil through decomposition of phosphate-rich organic compounds.

Solubilization of phosphate by plant roots and soil microorganisms is substantially influenced by various soil factors, such as PH, moisture and aeration.

In neutral or alkaline soils solubilization of phosphate is more as compared to acidic soils. Many phosphates solubilizing microorganisms are found in close proximity of root surfaces and may appreciably enhance phosphate assimilation by higher plants.

By their action, fungi bacteria and actinomycetes make available the organically bound phosphorus in soil and organic matter and the process is known as mineralization. On the other hand, certain microorganisms especially bacteria assimilate soluble phosphate and use for cell synthesis and on the death of bacteria, the phosphate is made available to plants. A fraction of phosphate is also lost in soil due to leaching. One of the ways to correct deficiency of phosphorus in plants is to inoculate seed or soil with commercial preparations (eg. Phosphobacterin) containing phosphate - solubilizing microorganisms along with phosphatic fertilizers.

Mineralization of phosphate is generally rapid and more in virgin soils than cultivated land. Mineralization is favoured by high temperatures (thermophilic range) and more in acidic to neutral soils with high organic phosphorus content. The enzyme involved in mineralization (cleavage) of phosphate from organic phosphorus compound is collectively called as "Phospatases".

The commercially used species of phosphate solubilizing bacteria and fungi are: *Bacillus polymyxa, Bacillus megatherium. Pseudomonas strita, Aspergillus, Penicllium avamori and Mycorrhiza*

Iron Cycle or Transformation

Iron exists in nature either as ferrous (Fe++) or ferric (Fe+++) ions. Ferrous iron is oxidized spontaneously to ferric state, forming highly insoluble ferric hydroxide. Plants as well as microorganisms require traces of iron, manganese copper, zinc, molybdenum, calcium boron, cobalt etc. Iron is always abundant in terrestrial habitats, and it is oftenly in an unavailable form for utilization by plants and leads to the serious deficiency in] plants.

Soil microorganisms play important role in the transformations of iron in al number of distinctly different ways such as:

A. Certain bacteria oxidize ferrous iron to ferric state which precipitate as ferric hydroxide around cells

B. Many heterotrophic species attack on in soluble organic iron salts and convert into inorganic salts

C. Oxidation-reduction potential decreases with microbial growth and which leads to the formation of more soluble ferrous iron from highly insoluble ferric ions

D. Number of bacteria and fungi produce acids such as carbonic, nitric, Sulphuric and organic acids which brings iron into solution

E. Under anaerobic conditions, the sulfides formed from sulphate and organic sulphur compounds remove the iron from solution as ferrous sulfide

F. As microbes liberate organic acids and other carbonaceous products of metabolism which results in the formation of soluble organic iron complex. Thus, iron may be precipitated in nature and immobilized by iron oxidizing bacteria under alkaline soil reaction and on the other hand solubilization of iron may occur through acid formation.

Some bacteria are capable of reducing ferric iron to ferrous which lowers the oxidation-reduction potential of the environment *(eg. Bacillus, Clostridium, Klebsiella* etc). However, some chemoautotrophic iron and sulphur bacteria such as *Thiobacillus ferroxldans* and *Ferrobacitlus ferrooxidans* can oxidize ferrous iron to ferric hydroxide which accumulates around the cells.

Most of the aerobic microorganisms live in an environment where iron exists in the oxidized, insoluble ferric hydroxide form. They produce iron-binding compounds in order to take up ferric iron. The iron-binding or chelating compounds / ligands produced by microorganisms are called "Siderophores". Bacterial siderophores may act as virulence factors in pathogenic bacteria and thus, bacteria that secrete siderophores are more virulent than non- siderophores producers. Therefore, siderophore producing bacteria can be used as biocontrol *agents eg. Fluorescent pseudomonads* used to control *Pythium,* causing damping-off diseases in seedlings. Recently Vascular - Arbusecular – Mycorrhiza (VAM) has been reported to increase uptake of iron.

Composition of Organic Matter

Soil organic matter plays important role in the maintenance and improvement of soil properties. It is a dynamic material and is one of the major sources of nutrient elements for plants. Soil organic matter is derived to a large extent from residues and remains of the plants together with the small quantities of animal remains, excreta, and microbial tissues. Soil organic matter is composed of three major components i.e. plants residues, animal remain and dead remains of microorganisms. Various organic compounds are made up of complex carbohydrates, (Cellulose, hemicellulose, starch) simple sugars, lignins, pectins, gums, mucilages, proteins, fats, oils, waxes, resins, alcohols, organic acids, phenols etc. and other products. All these compounds constituting the soil organic matter can be categorized in the following way.

Organic Matter (Undecomposed)

Organic:

- Nitrogenous:
    1. Water Soluble eg. Nitrates, ammonical compounds, amides, amino acids etc.
    2. Insoluble eg. Proteins nucleoproteins, peptides, alkaloids purines, pyridines chitin etc.

- Non Nitrogenous:
  - o Carbohydrates eg. Sugars, starch, hemicellulose, gums, mucilage, pectins, etc.
  - o Micellaneous: eg. Lignin, tannins, organic acid, etc.
  - o Ether Solube: eg. Fats, oils, wax etc.

Inorganic: The organic complex / matter in the soil is, therefore made up of a large number of substances of widely different chemical composition and the amount of each substance varies with the type, nature and age of plants. For example cellulose in a young plant is only half of the mature plants; water-soluble organic substances in young plants are nearly double to that of older plants. Among the plant residues, leguminous plants are rich in proteins than the non-leguminous plants. Grasses and cereal straws contain greater amount of cellulose, lignin, hemicelluloses than the legumes and as the plant gets older the proportion of cellulose, hemicelluloses and lignin gets increased. Plant residues contain 15-60% cellulose, 10-30 % hemicelluslose, 5-30% lignin, 2-15 % protein and 10% sugars, amino acids and organic acids. These differences in composition of various plant and animal residues have great significance on the rate of organic matter decomposition in general and of nitrification and humification (humus formation) in particular. The end products of decomposition are CO2, H2O, NO3, SO4, CH4, NH4, and H2S depending on the availability of air.

Factors Influencing Rate of Organic Matter Decomposition

In addition to the composition of organic matter, nature and abundance of microorganisms in soil, the extent of C, N, P and K., moisture content of the soil and its temperature, PH, aeration, C: N ratio of plant residues and presence/absence of inhibitory substances (e.g. tannins) etc. are some of the major factors which influence the rate of organic matter decomposition.

As soon as plant and animal residues are added to the soil, there is a rapid increase in the activity of microorganisms. These are not true soil organisms, but they continue their activity by taking part in the decomposition of organic matter and thereby release of plant nutrients in the soil. Bacteria are the most abundant organisms playing important role in the decomposition of organic matter. Majority of bacteria involved in decomposition of organic matter are heterotrophs and autotrophs are least in proportion which are not directly involved in organic matter decomposition. Actinomycetes and fungi are also found to play important role in the decomposition of organic matter.

Soil algae may contribute a small amount of organic matter through their biomass but they do not have any active role in organic matter decomposition. The various microorganisms involved in the decomposition of organic matter are listed in the following table.

	Microorganisms		
Constituents	***Bacteria***	***Fungi***	***Actinomycetes***
Cellulose	*Achromobacter, Bacillus, Cellulomonas, Cellvibrio, Clostridium, Cytophaga, Vibrio Pseudomonas, Sporocytophaga etc.*	*Aspergillus, Chaetomium, Fusarium, Pencillium Rhizoctonia, Rhizopus, Trichoderma, Verticillttm.*	*Micromonospora, Nocardia Streptomyces, Thermonospora*
Hemicellulose	*Bacillus, Achromobacter, Cytophaga Pseudomonas, Erwinia, Vibrio, Lactobacillus*	*Aspergillus, Fusarium, Chaetomium, Penicillium, Trichoderma, Humicola*	*Streptomyces, Actinomycetes*
Lignin	*Flavobacterium, Pseudomonas, Micrococcus, Arthorbacter, Xanthomonas*	*Humicola, Fusarium Fames, Pencillium, Aspergillus, Ganoderma*	*Streptomyces, Nocardia*
Starch	*Achromobacter, Bacillus, Clostridium*	*Fusarium, Fomes, Aspergillus, Rhizopus*	*Micromonospora, Nocardia, Streptomyces,*
Pectin	*Bacillus, Clostridium, Pseudomonas*	*Ftisarium, Verticillum*	
Chitin	*Bacillus, Achromobacter, Cytophaga, Pseudomonas*	*Mucor, Fusarium, Aspergillus, Trichoderma*	*Streptomyces, Nocardia, Micromanospora*
Proteins & Nucleic acids	*Bacillus, Pseudomonas, Clostriddum, Serratia, Micrococcus*	*Penicillium, Rhodotorula,*	*Streptomyces*

Aeration: Good aeration is necessary for the proper activity of the microorganisms involved in the decomposition of organic matter. Under anaerobic conditions fungi and actinomycetes are almost suppressed and only a few bacteria *(Clostridium)* take part in anaerobic decomposition. The rate of decomposition is markedly retarded. It was found that under aerobic conditions 65 percent of the total organic matter decomposes during six months, while under anaerobic conditions only 47 percent organic matter can be decomposed during the same period. Anaerobic decomposition of organic matter results into the production of large quantity of organic acids and evolution of gases like methane (CH 4) hydrogen (H2) and carbon dioxide (CO2).

Temperature: The rate of decomposition is more rapid in the temperature range of 30° to 40°' At temperatures below or above this range, the rate of decomposition is markedly retarded. Appreciable organic mater decomposition occurs at 25° C and further fluctuation in the soil temperature has little effect on decomposition.

Moisture: Adequate soil moisture i.e. about 60 to 80 percent of the water-holding capacity of the soil is must for the proper decomposition of organic matter. Too much moisture leads to insufficient aeration which results in the reduced activity of microorganisms and there by checks the rate of decomposition.

Soil PH/soil Reaction: Soil PH affects directly the kind, density and the activity of fungi, bacteria & actinomycetes involved in the process of decomposition and thereby rate of decomposition of organic matter. The rate of decomposition is more in neutral soils than that of acidic soils. Therefore, treatment of acid soils with lime can accelerate the rate of organic matter decomposition.

C: N Ratio: C: N ration of organic matter has great influence on the rate of decomposition. Organic matter from diverse plant-tissues varies widely in their C: N ratio (app. 8-10 %). The optimum C: N ratio in the range of 20-25 is ideal for maximum decomposition, since a favourable soil environment is created to bring about equilibrium between mineralization and immobilization processes. Thus, a low nitrogen content or wide C'.N ratio results into the slow decomposition. Protein rich, young and succulent plant tissues are decomposed more rapidly than die protein-poor, mature and hard plant tissues. Therefore, C:N ratio of organic matter as well as soil should be narrow/less for better and rapid decomposition. Thus, high aeration, mesophilic temperature range, optimum moisture, neutral/alkaline soil reaction and narrow C: N ratio of soil and organic matter are required for rapid and better decomposition of organic matter.

Microbiology of Decomposition of Various Constituents in Organic Matter

When plant and animal residues are added to the soil, the various constituents of the soil organic matter are decomposed simultaneously by the activity of microorganisms and carbon is released as CO2, and nitrogen as NH4 —>NO3 for the use by plants. Other nutrients are also converted into plant usable forms. This process of release of nutrients from organic matter is called mineralization. The insoluble plant residues constitute the part of humus and soil organic matter complex. The final product of aerobic decomposition is CO2 and that of anaerobic decomposition are Hydrogen, ethyl alcohol (CH4), various organic acids and carbon dioxide (CO2). Soil organisms use organic matter as a source of energy and food.

The process of decomposition is initially fast, but slows down considerably as the supply of readily decomposable organic matter gets exhausted. Sugars, water-soluble nitrogenous compounds, amino acids, lipids, starches and some of the hemicellulases are decomposed first at rapid rate, while insoluble compounds such as cellulose, hemicellulose, lignin, proteins etc. which forms the major portion of organic matter are decomposed later slowly. Thus, the organic matter added to the soil is converted by oxidative decomposition to simpler substances which are made available in stages for plant growth and the residue is transformed into humus.

The microbiology of decomposition/degradation of some of the major constituents (viz. Cellulose, Hemicellulose, Lignin, Proteins etc.) of soil organic matter/plant residues are discussed in brief in the following paragraphs.

Decomposition of Cellulose: Cellulose is the most abundant carbohydrate present in plant residues/organic matter in nature. When cellulose is associated with pentosans (eg. xylans & mannans) it undergoes rapid decomposition, but when associated with lignin, the rate of decomposition is very slow. The decomposition of cellulose occurs in two stages: (i) in the first stage the long chain of cellulase is broken down into cellobiase and then into glucose by the process of hydrolysis in the presence of enzymes cellulase and cellobiase, and (ii) in second stage glucose is oxidized and converted CO2 and water.

Cellulase Cellobiase

1. Cellulose ————> Cellobiose ————> Glucose

Hydrolysis hydrolysis

Oxidation Oxidation

2. Glucose ————> Organic Acids ————> CO2 + H2O

The intermediate products formed/released during enzymatic hydrolysis of cellulose (eg. cellobiose and glucose) are utilized by the cellulose-decomposing organisms or by other organisms as source of energy for biosynthetic processes. The cellulolytic microorganisms responsible for degradation of cellulose through the excretion of enzymes (cellulase & Cellobiase) are fungi, bacteria and actinomycetes.

Decomposition of Hemicelluloses: Hemicelluloses are water-soluble polysaccharides and consists of hexoses, pentoses, and uronic acids and are the major plant constituents second only in quantity of cellulose, and sources of energy and nutrients for soil microflora.

When subjected to microbial decomposition, hemicelluloses degrade initially at faster rate and are first hydrolyzed to their component sugars and uronic acids. The hydrolysis is brought about by number of hemicellulolytic enzymes known as "hemicellulases" excreted by the microorganisms. On hydrolysis hemicelluloses are converted into soluble monosaccharide/sugars (eg. xylose, arabinose, galactose and mannose) which are further convened to organic acids, alcohols, CO2 and H2O and uronic acids are broken down to pentoses and CO2. Various microorganisms including fungi, bacteria and actinomycetes both aerobic and anaerobic are involved in the decomposition of hemicelluloses.

Lignin Decomposition: Lignin is the third most abundant constituent of plant tissues, and accounts about 10-30 percent of the dry matter of mature plant materials. Lignin content of young plants is low and gradually increases as the plant grows old. It is one of the most resistant organic substances for the microorganisms to degrade however certain Basidiomycetous fungi are known to degrade lignin at slow rates. Complete oxidation of lignin result in the formation of aromatic compounds such as syringaldehydes, vanillin and ferulic acid. The final cleavages of these aromatic compounds yield organic acids, carbon dioxide, methane and water.

Protein Decomposition: Proteins are complex organic substances containing nitrogen, sulphur, and sometimes phosphorus in addition to carbon, hydrogen and oxygen. During the course of decomposition of organic matter, proteins are first hydrolyzed to a number of intermediate products eg. Proteases, peptides etc. collectively known as polypeptides

The intermediate products so formed are then hydrolyzed and broken down ultimately to individual amino acids, or ammonia and amides. The process of hydrolysis of proteins to amino acids is known as "aminization or ammonification", which is brought about by certain enzymes, collectively known as "proteases" or "proteolytic" enzymes secreted by various microorganisms. Amino acids and amines are further decomposed and converted into ammonia. During the course of ammonification, various organic acids, alcohols, aldehydes etc. are produced which are further decomposed finally to produce carbon dioxide and water.

All types of microorganisms, bacteria, fungi, and actinomycetes are able to bring about decomposition of proteins. In acid soils, fungi are pre-dominant, while in neutral and alkaline soils bacteria are dominant decomposers of proteins.

Ecological Association/Interactions Among Soil Microorganisms

Soil is the largest terrestrial ecosystem where a wide variety of relationships exists between different types of soil organisms. The associations existing between different soil microorganisms, whether of a symbiotic or antagonistic nature, influence the activities of microorganisms in the soil. Microflora composition of any habitat is governed by the biological equilibrium created by the associations and interactions of all individuals found in the community. In soil, many microorganisms live in close proximity and interact among themselves in a different ways. Some of the interactions or associations are mutually beneficial, or mutually detrimental or neutral. The various types of possible interactions/associations occurring among the microorganisms in soil can be: a) beneficial i) mutualism ii) commensalisms and iii) proto-cooperation or b) detrimental / harmful - i) amensalism, ii) antagonism, iii) competition iv) Parasitism and v) predation

Beneficial Association/Interactions

Mutualism (Symbiosis): It is a relationship or a type of symbiosis in which both the interacting organisms/partners are benefited from each other. The way/manner in which benefit is derived depends on the type of interactions. When the benefit is in the term of exchange of nutrients, then the relationship is termed as "syntrophism" (Greek meaning: Syn -mutual and trophe = nourishment), for example, Lichen (association of algae or BGA with fungus) in which algae benefits by protection afforded to it by the fungal hyphae from environmental stresses, while the fungus obtain and use CO2 released by the algae during photosynthesis. Where the blue green algae are the partners in the lichen association, the heterotrophs (Fungus), benefit from the fixed nitrogen by the blue green algae.

Microorganisms may also form mutualistic relationships with plants, for example nitrogen fixing bacteria i.e. *Rhizobium* growing in the roots of legumes. In this Rhizobium-legume association, *Rhizobium* bacteria are benefited by protection from the environmental stresses while in turn plant is benefited by getting readily available nitrate nitrogen released by the bacterial partner.

The *Anabaena-Azolla* is an association between the water fern *Azolla* and the cyanobacterium Amabaena. This association is of great importance in paddy fields, where nitrogen is frequently a limiting nutrient.

An actinorrhizal symbiosis of actinomycetes, *Frankia* with the roots of *Alnus* and *Casurina (non*-legumes) is common in temperate forest ecosystem for soil nitrogen economy. Another type of symbiotic association which exists between the roots of higher plants and fungus is *Mycorrhiza.* In this association fungus gets essential organic nutrients and protection form roots of the plants and allows them to multiply and in turn plants uptake phosphorus, nitrogen and other inorganic nutrients made available by the fungus.

Commensalisms

In this association one organism/partner in association is benefited by other partner without affecting it. For example, many fungi can degrade cellulose to glucose, which is utilized by many bacteria. Lignin which is major constituent of woody plants and is usually resistant to degradation by most of the microorganisms but in forest soils, lignin is readily degraded by a group of Basidiomycetous fungi and the degraded products are used by several other fungi and bacteria which can not utilize lignin directly. This type of association is also found in organic matter decomposition process.

Proto-cooperation

It is mutually beneficial association between two species / partners. Unlike symbiosis, proto-cooperation is not obligatory for their existence or performance of a particular activity. In this type of association one organism favour its associate by removing toxic substances from the habitat and simultaneously obtain carbon products made by the another associate/partner. Nutritional proto-cooperation between bacteria and fungi has been reported for various vitamins, amino and purines in terrestrial ecosystem and are very useful in agriculture. Proto-cooperative associations found beneficial in agriculture are : i) synergism between VAM fungus-legume plants and *Rhizobium* in which nitrogen fixation and phosphorus availability / uptake is much higher resulting in higher crop yields and improved soil fertility, ii) synergism between PSM-legume plants and *Rhizobium* and iii) synergism between plant roots and PGPR in rhizosphere where rhizobacteria restrict the growth of phytopathogens on plant roots and secrets growth promoting substances.

Detrimental (Harmful) Associations/Interactions:

Antagonism: It is the relationship in which one species of an organism is inhibited or adversely affected by another species in the same environment. In such antagonism, one organism may directly or indirectly inhibit the activities of the other. Antagonistic relations are most common in nature and are also important for the production of antibiotics. The phenomenon of antagonism may be categorized into three i.e. antibiosis, competition and exploitation.

In the process of antibiosis, the antibiotics or metabolites produced by one organism inhibits another organism. An antibiotic is a microbial inhibitor of biological origin. Innumerable examples of antibiosis are found in soil. For example, *Bacillus* Species from soil produces an antifungal agent which inhibits growth of several soil fungi. Several species of *Streptomyces* from soil produces antibacterial and antifungal antibiotics. Most of the commercial antibiotics such as streptomycin, chloramphenicol, Terramycin and cyclohexamide have been produced from the mass culture of *Streptomyces.* Thus, species of *Streptomyces* are the largest group of antibiotic producer's in soil. Another example of antibiosis is inhibition of *Verticillium* by *Trichoderma,* inhibition of *Rhizoctonia* by a bacterium *Bacillus subtilis,* inhibition of soil fungus *Aspergillus terreus* by a bacterium *Staphylococcus aureus.*

Ammensalism: In this interaction /association one partner suppress the growth of other partner by producing toxins like antibiotics and harmful gases like ethylene, HCN, Nitrite etc.

Competition: As soil, is inhabited by many different species of microorganisms, there exists an active competition among them for available nutrients and space. The limiting substrate may result in favouring one species over another. Thus, competition can be defined as "the injurious effect of one organism on another because of the removal of some resource of the environment". This phenomenon can result in major fluctuations in the composition of the microbial population in the soil.

For example, chlamydospores of *Fusarium, Oospores of* Aphanomyces and conidia of *Verticillium dahlae* require exogenous nutrients to germinate in soil. But other fungi and soil bacteria deplete these critical nutrients required for spore germination and thereby hinder the spore germination resulting into the decrease in population. Competition for free space has been, reported to suppress the fungal population by soil bacteria. Therefore, organisms with inherent ability to grow fast are better competitors.

Parasitism: It is an association, in which one organism lives in or on the body of another. The parasite is dependent upon the host and lives in intimate physical contact and forms metabolic association with the host. So this is a host -parasite relationship in which one (parasite) is benefited while other (host) is adversely affected, although not necessarily killed. Parasitism is widely spread in soil communities, for example, bacteriophages (viruses which attack bacteria) are strict intracellular parasites Chytrid fungi, which parasitize algae, as well as other fungi and plants; there are many strains of fungi which are parasitic on algae, plants, animals parasitized by different organisms, earthworms are parasitized by fungi, bacteria, viruses etc.

Predation: Predation is an association / exploitation in which predator organism directly feed on and kills the pray organism. It is one of the most dramatic inter relationship among microorganisms in nature, for example, the nematophagous fungi are the best examples of predatory soil fungi. Species of *Arthrobotrytis* and *Dactylella* are known as nematode trapping fungi. Other examples of microbial predators are the protozoa and slime mold fungi which feed on the bacteria and reduce their population. The bacteriophages may also be considered as predators of bacteria.

Soil Microorganisms in Biodegradation of Pesticides and Herbicides

Pesticides are the chemical substances that kill pests and herbicides are the chemicals that kill weeds. In the context of soil, pests are fungi, bacteria insects, worms, and nematodes etc. that cause damage to field crops. Thus, in broad sense pesticides are insecticides, fungicides, bactericides, herbicides and nematicides that are used to control or inhibit plant diseases and insect pests. Although wide-scale application of pesticides and herbicides is an essential part of augmenting crop yields; excessive use of these chemicals leads to the microbial imbalance, environmental pollution and health hazards. An ideal pesticide should have the ability to destroy target pest quickly and should be able to degrade non-toxic substances as quickly as possible.

The ultimate "sink" of the pesticides applied in agriculture and public health care is soil. Soil being the storehouse of multitudes of microbes, in quantity and quality, receives the chemicals in various forms and acts as a scavenger of harmful substances. The efficiency and the competence to handle the chemicals vary with the soil and its physical, chemical and biological characteristics.

Effects of Pesticides

Pesticides reaching the soil in significant quantities have direct effect on soil microbiological aspects, which in turn influence plant growth. Some of the most important effects caused by pesticides are : (1) alterations hi ecological balance of the soil microflora, (2) continued application of large quantities of pesticides may cause ever lasting changes in the soil microflora, (3) adverse effect on soil fertility and crop productivity, (4) inhibition of N2 fixing soil microorganisms such as *Rhizobium, Azotobacter, Azospirillum* etc. and cellulolytic and phosphate solubilizing microorganisms, (5) suppression of nitrifying bacteria, *Nitrosomonas* and *Nitrobacter* by soil fumigants ethylene bromide, Telone, and vapam have also been reported, (6) alterations in nitrogen balance of the soil, (7) interference with ammonification in soil, (8) adverse effect on mycorrhizal symbioses in plants and nodulation in legumes, and (9) alterations in the rhizosphere microflora, both quantitatively and qualitatively.

Persistence of Pesticides in Soil

How long an insecticide, fungicide, or herbicide persists in soil is of great importance in relation to pest management and environmental pollution. Persistence of pesticides in soil for longer period is undesirable because of the reasons: a) accumulation of the chemicals in soil to highly toxic levels, b) may be assimilated by the plants and get accumulated in edible plant products, c) accumulation in the edible portions of the root crops, d) to be get eroded with soil particles and may enter into the water streams, and finally leading to the soil, water and air pollutions. The effective persistence of pesticides in soil varies from a week to several years depending upon structure and properties of the constituents in the pesticide and availability of moisture in soil. For instance, the highly toxic phosphates do not persist for more than three months while chlorinated hydrocarbon insecticides (eg. DOT, aldrin, chlordane etc) are known to persist at least for 4-5 years and some times more than 15 years.

From the agricultural point of view, longer persistence of pesticides leading to accumulation of residues in soil may result into the increased absorption of such toxic chemicals by plants to the level at which the consumption of plant products may prove deleterious / hazardous to human beings as well as livestock's. There is a chronic problem of agricultural chemicals, having entered in food chain at highly inadmissible levels in India, Pakistan, Bangladesh and several other developing countries in the world. For example, intensive use of DDT to control insect pests and mercurial fungicides to control diseases in agriculture had been known to persist for longer period and thereby got accumulated in the food chain leading to food contamination and health hazards. Therefore, DDT and mercurial fungicides has been, banned to use in agriculture as well as in public health department.

Biodegradation of Pesticides in Soil

Pesticides reaching to the soil are acted upon by several physical, chemical, and biological forces. However, physical and chemical forces are acting upon/degrading the pesticides to some extent, microorganism's plays major role in the degradation of pesticides. Many soil microorganisms have the ability to act upon pesticides and convert them into simpler non-toxic compounds. This process of degradation of pesticides and conversion into non-toxic compounds by microorganisms is known as "biodegradation". Not all pesticides reaching to the soil are biodegradable and such chemicals that show complete resistance to biodegradation are called "recalcitrant".

The chemical reactions leading to biodegradation of pesticides fall into several broad categories which are discussed in brief in the following paragraphs.

Detoxification

Conversion of the pesticide molecule to a non-toxic compound. Detoxification is not synonymous with degradation. Since a single chance in the side chain of a complex molecule may render the chemical non-toxic.

Degradation

The breaking down / transformation of a complex substrate into simpler products leading finally to mineralization. Degradation is often considered to be synonymous with mineralization, e.g. Thirum (fungicide) is degraded by a strain of *Pseudomonas* and the degradation products are dimethlamine, proteins, sulpholipaids, etc.

Conjugation (Complex Formation or Addition Reaction)

In which an organism make the substrate more complex or combines the pesticide with cell metabolites. Conjugation or the formation of addition product is accomplished by those organisms catalyzing the reaction of addition of an amino acid, organic acid or methyl crown to the substrate, for e.g., in the microbial metabolism of sodium dimethly dithiocarbamate, the organism combines the fungicide with an amino acid molecule normally present in the cell and thereby inactivate the pesticides/chemical.

Activation

It is the conversion of non-toxic substrate into a toxic molecule, for eg. Herbicide, 4-butyric acid (2, 4-D B) and the insecticide Phorate are transformed and activated microbiologically in soil to give metabolites that are toxic to weeds and insects.

Changing the Spectrum of Toxicity

Some fungicides/pesticides are designed to control one particular group of organisms / pests, but they are metabolized to yield products inhibitory to entirely dissimilar groups of organisms, for e.g. the fungicide PCNB fungicide is converted in soil to chlorinated benzoic acids that kill plants.

Biodegradation of pesticides / herbicides is greatly influenced by the soil factors like moisture, temperature, PH and organic matter content, in addition to microbial population and pesticide solubility. Optimum temperature, moisture and organic matter in soil provide congenial environment for the break down or retention of any pesticide added in the soil. Most of the organic pesticides degrade within a short period (3-6 months) under tropical conditions. Metabolic activities of bacteria, fungi and actinomycetes have the significant role in the degradation of pesticides.

Criteria for Bioremediation / Biodegradation

For successful biodegradation of pesticide in soil, following aspects must be taken into consideration. i) Organisms must have necessary catabolic activity required for degradation of contaminant at fast rate to bring down the concentration of contaminant, ii) the target contaminant must be bioavailability, iii) soil conditions must be congenial for microbial /plant growth and enzymatic activity and iv) cost of bioremediation must be less than other technologies of removal of contaminants.

According to Gales (1952) principal of microbial infallibility, for every naturally occurring organic compound there is a microbe / enzyme system capable its degradation.

Strategies for Bioremediation

For the successful biodegradation / bioremediation of a given contaminant following strategies are needed.

a) Passive/ intrinsic Bioremediation: It is the natural bioremediation of contaminant by tile indigenous microorganisms and the rate of degradation is very slow.

b) Biostimulation: Practice of addition of nitrogen and phosphorus to stimulate indigenous microorganisms in soil.

c) Bioventing: Process/way of Biostimulation by which gases stimulants like oxygen and methane are added or forced into soil to stimulate microbial activity.

d) Bioaugmentation: It is the inoculation/introduction of microorganisms in the contaminated site/soil to facilitate biodegradation.

e) Composting: Piles of contaminated soils are constructed and treated with aerobic thermophilic microorganisms to degrade contaminants. Periodic physical mixing and moistening of piles are done to promote microbial activity.

f) Phytoremediation: Can be achieved directly by planting plants which hyperaccumulate heavy metals or indirectly by plants stimulating microorganisms in the rhizosphere.

g) Bioremediation:Process of detoxification of toxic/unwanted chemicals / contaminants in the soil and other environment by using microorganisms.

h) Mineralization: Complete conversion of an organic contaminant to its inorganic constituent by a species or group of microorganisms.

Introduction to Biofertilizers

Biofertilizers are microbial inoculants or carrier based preparations containing living or latent cells of efficient strains of nitrogen fixing, phosphate is solublizing and cellulose decomposing microorganisms intended for seed or soil application and designed to improve soil fertility and plant growth by increasing the number and biological activity of beneficial microorganisms in the soil.

The objects behind the application of Biofertilizers /microbial inoculants to seed, soil or compost pit is to increase the number and biological / metabolic activity of useful microorganisms that accelerate certain microbial processes to augment the extent of availability of nutrients in the available forms which can be easily assimilated by plants. The need for the use of Biofertilizers has arisen primarily due to two reasons i.e. though chemical fertilizers increase soil fertility, crop productivity and production, but increased / intensive use of chemical fertilizers has caused serious concern of soil texture, soil fertility and other environmental problems, use of Biofertilizers is both economical as well as environment friendly. Therefore, an integrated approach of applying both chemical fertilizers and Biofertilizers is the best way of integrated nutrient supply in agriculture.

Organic fertilizers (manure, compost, vermicompost) are also considered as Biofertilizers, which are rendered in available forms due to the interactions of microorganisms or their association with plants. Biofertilizers, thus include i) Symbiotic nitrogen fixers *Rhizobium sp.* ii) Non-symbiotic, free living nitrogen fixers *Azotobacter, Azospirillum* etc. iii) BGA-inoculants *Azolla-Anabaena,* iv) Phosphate solubilizing microorganisms (PSM) *Bacillus Pseudomonas, Penicillium Aspergillus* etc. v) Mycorrhiza vi) Cellulolytic microorganisms and vii) Organic fertilizers.

Nobbe and Hiltner (1895, USA) produced the first *Rhizobium* biofertilizer under the brand name "Nitragin" for 17 different legumes.

Role of Biofertilizers in Soil Fertility and Agriculture

Biofertilizers are known to play a number of vital roles in soil fertility, crop productivity and production in agriculture as they are eco friendly and can not at any cost replace chemical fertilizers that are indispensable for getting maximum crop yields. Some of the important functions or roles of Biofertilizers in agriculture are:

1. They supplement chemical fertilizers for meeting the integrated nutrient demand of the crops.
2. They can add 20-200 kg N/ha year *(eg. Rhizobium sp* 50-100 kg N/ha year ; *Azospirillum , Azotobacter* : 20-40 kg N/ha /yr; Azolla : 40-80 kg N/ha; BGA :20-30 kg N/ha) under optimum soil conditions and thereby increases 15-25 percent of total crop yield.
3. They can at best minimize the use of chemical fertilizers not exceeding 40-50 kg N/ha under ideal agronomic and pest-free conditions.
4. Application of Biofertilizers results in increased mineral and water uptake, root development, vegetative growth and nitrogen fixation.
5. Some Biofertilizers *(eg, Rhizobium BGA, Azotobacter* sp) stimulate production of growth promoting substance like vitamin-B complex, Indole acetic acid (IAA) and Gibberellic acids etc.
6. Phosphate mobilizing or phosphorus solubilizing Biofertilizers / microorganisms (bacteria, fungi, mycorrhiza etc.) converts insoluble soil phosphate into soluble forms by secreting several organic acids and under optimum conditions they can solubilize / mobilize about 30-50 kg P2O5/ha due to which crop yield may increase by 10 to 20%.
7. Mycorrhiza or VA-mycorrhiza (VAM fungi) when used as Biofertilizers enhance uptake of P, Zn, S and water, leading to

uniform crop growth and increased yield and also enhance resistance to root diseases and improve hardiness of transplant stock.

8. They liberate growth promoting substances and vitamins and help to maintain soil fertility.
9. They act as antagonists and suppress the incidence of soil borne plant pathogens and thus, help in the bio-control of diseases.
10. Nitrogen fixing, phosphate mobilizing and cellulolytic microorganisms in bio-fertilizer enhance the availability of plant nutrients in the soil and thus, sustain the agricultural production and farming system.
11. They are cheaper, pollution free and renewable energy sources
12. They improve physical properties of soil, soil tilth and soil health in general.
13. They improve soil fertility and soil productivity.
14. Blue green algae like *Nostoc, Anabaena, and Scytonema* are often employed in the reclamation of alkaline soils.
15. Bio-inoculants containing cellulolytic and lignolytic microorganisms enhance the degradation/ decomposition of organic matter in soil, as well as enhance the rate of decomposition in compost pit.
16. BGA plays a vital role in the nitrogen economy of rice fields in tropical regions.
17. *Azotobacter* inoculants when applied to many non-leguminous crop plants, promote seed germination and initial vigor of plants by producing growth promoting substances.
18. *Azolla-Anabaena* grows profusely as a floating plant in the flooded rice fields and can fix 100-150 kg N/ha /year in approximately 40-60 tones of biomass produced,
19. Plays important role in the recycling of plant nutrients.

Quality Control Measures (as per ISI Specifications)

1. Since, Biofertilizers contains live cells, care should be taken during their transportation and storage.
2. They should be kept in a cold place and not exposed to sunlight.
3. Biofertilizers for legumes are crop-specific; therefore, they must be used for the crop for which they are meant.

4. Biofertilizers when used under adverse soil conditions, appropriate remedial measures (liming and use of Gypsum) should be followed.
5. Biofertilizers must be carrier-based
6. Carrier material used should be in form of powder (75-106 micron size.
7. It should contain minimum of 10^8 viable cells of microorganisms /gram of the carrier material on dry weight basis.
8. It should have a minimum period of six months expiry from date of its
9. It should be tree from any contaminant /contamination with other microorganisms.
10. PH should be in the range of 6.0-7.5.
11. It should induce desired beneficial effects on all those crops, species /cultivars listed on the packet before the expiry date.
12. It should be packed in 50-75 micron low density polythene packets

4

Soil Enzymes

Catalase

Catalase is a common enzyme found in nearly all living organisms exposed to oxygen. It catalyzes the decomposition of hydrogen peroxide to water and oxygen. It is a very important enzyme in protecting the cell from oxidative damage by reactive oxygen species (ROS). Likewise, catalase has one of the highest turnover numbers of all enzymes; one catalase molecule can convert millions of molecules of hydrogen peroxide to water and oxygen each second.

Catalase is a tetramer of four polypeptide chains, each over 500 amino acids long. It contains four porphyrin heme (iron) groups that allow the enzyme to react with the hydrogen peroxide. The optimum pH for human catalase is approximately 7, and has a fairly broad maximum (the rate of reaction does not change appreciably at pHs between 6.8 and 7.5). The pH optimum for other catalases varies between 4 and 11 depending on the species. The optimum temperature also varies by species.

History

Catalase was first noticed in 1811 when Louis Jacques Thénard, who discovered H_2O_2 (hydrogen peroxide), suggested its breakdown is caused by an unknown substance. In 1900, Oscar Loew was the first to give it the name catalase, and found it in many plants and animals. In 1937 catalase from beef liver was crystallised by James B. Sumner and Alexander Dounce and the molecular weight was found in 1938.

In 1969, the amino acid sequence of bovine catalase was discovered. Then in 1981, the three-dimensional structure of the protein was revealed.

Action

The reaction of catalase in the decomposition of living tissue:

$$2\ H_2O_2 \rightarrow 2\ H_2O + O_2$$

The presence of catalase in a microbial or tissue sample can be tested by adding a volume of hydrogen peroxide and observing the reaction. The formation of bubbles, oxygen, indicates a positive result. This easy assay, which can be seen with the naked eye, without the aid of instruments, is possible because catalase has a very high specific activity, which produces a detectable response.

Molecular Mechanism

While the complete mechanism of catalase is not currently known, the reaction is believed to occur in two stages:

$$H_2O_2 + Fe(III)\text{-}E \rightarrow H_2O + O{=}Fe(IV)\text{-}E(.+)$$

$$H_2O_2 + O{=}Fe(IV)\text{-}E(.+) \rightarrow H_2O + Fe(III)\text{-}E + O_2$$

Here Fe()-E represents the iron centre of the heme group attached to the enzyme. Fe(IV)-E(.+) is a mesomeric form of Fe(V)-E, meaning the iron is not completely oxidized to +V, but receives some "supporting electrons" from the heme ligand. This heme has to be drawn then as a radical cation (.+).

As hydrogen peroxide enters the active site, it interacts with the amino acids Asn147 (asparagine at position 147) and His74, causing a proton (hydrogen ion) to transfer between the oxygen atoms. The free oxygen atom coordinates, freeing the newly formed water molecule and Fe(IV)=O. Fe(IV)=O reacts with a second hydrogen peroxide molecule to reform Fe(III)-E and produce water and oxygen. The reactivity of the iron centre may be improved by the presence of the phenolate ligand of Tyr357 in the fifth iron ligand, which can assist in the oxidation of the Fe(III) to Fe(IV). The efficiency of the reaction may also be improved by the interactions of His74 and Asn147 with reaction intermediates. In general, the rate of the reaction can be determined by the Michaelis-Menten equation.

Catalase can also catalyze the oxidation, by hydrogen peroxide, of various metabolites and toxins, including formaldehyde, formic acid, phenols, acetaldehyde and alcohols. It does so according to the following reaction:

$$H_2O_2 + H_2R \rightarrow 2H_2O + R$$

The exact mechanism of this reaction is not known.

Any heavy metal ion (such as copper cations in copper(II) sulphate) can act as a noncompetitive inhibitor of catalase. Also, the poison cyanide is a competitive inhibitor of catalase, strongly binding to the heme of catalase and stopping the enzyme's action.

Three-dimensional protein structures of the peroxidated catalase intermediates are available at the Protein Data Bank. This enzyme is commonly used in laboratories as a tool for learning the effect of enzymes upon reaction rates.

Cellular Role

Hydrogen peroxide is a harmful byproduct of many normal metabolic processes; to prevent damage to cells and tissues, it must be quickly converted into other, less dangerous substances. To this end, catalase is frequently used by cells to rapidly catalyze the decomposition of hydrogen peroxide into less-reactive gaseous oxygen and water molecules.

The true biological significance of catalase is not always straightforward to assess: Mice genetically engineered to lack catalase are phenotypically normal, indicating this enzyme is dispensable in animals under some conditions. A catalase deficiency may increase the likelihood of developing type 2 diabetes. Some humans have very low levels of catalase (acatalasia), yet show few ill effects. The predominant scavengers of H_2O_2 in normal mammalian cells are likely peroxiredoxins rather than catalase.

Human catalase works at an optimum temperature of 45°C, which is approximately the temperature of the human body. In contrast, catalase isolated from the hyperthermophile archaeon *Pyrobaculum calidifontis* has a temperature optimum of 90°C.

Catalase is usually located in a cellular, bipolar environment organelle called the peroxisome. Peroxisomes in plant cells are involved in photorespiration (the use of oxygen and production of carbon dioxide) and symbiotic nitrogen fixation (the breaking apart of diatomic nitrogen (N_2) to reactive nitrogen atoms). Hydrogen peroxide is used as a potent antimicrobial agent when cells are infected with a pathogen. Catalase-positive pathogens, such as *Mycobacterium tuberculosis*, *Legionella pneumophila*, and *Campylobacter jejuni*, make catalase to deactivate the peroxide radicals, thus allowing them to survive unharmed within the host.

Catalase contributes to ethanol metabolism in the body after ingestion of alcohol, but it only breaks down a small fraction of the alcohol in the body.

Distribution Among Organisms

All known animals use catalase in every organ, with particularly high concentrations occurring in the liver. One unique use of catalase occurs in the bombardier beetle. This beetle has two sets of chemicals ordinarily stored separately in its paired glands. The larger of the pair, the storage chamber or reservoir, contains hydroquinones and hydrogen peroxide, whereas the smaller of the pair, the reaction chamber, contains catalases and peroxidases. To activate the noxious spray, the beetle mixes the contents of the two compartments, causing oxygen to be liberated from hydrogen peroxide. The oxygen oxidizes the hydroquinones and also acts as the propellant. The oxidation reaction is very exothermic (ΔH = "202.8 kJ/mol) which rapidly heats the mixture to the boiling point.

Catalase is also universal among plants, and many fungi are also high producers of the enzyme.

Almost all aerobic microorganisms use catalase. It is also present in some anaerobic microorganisms, such as *Methanosarcina barkeri*.

Applications

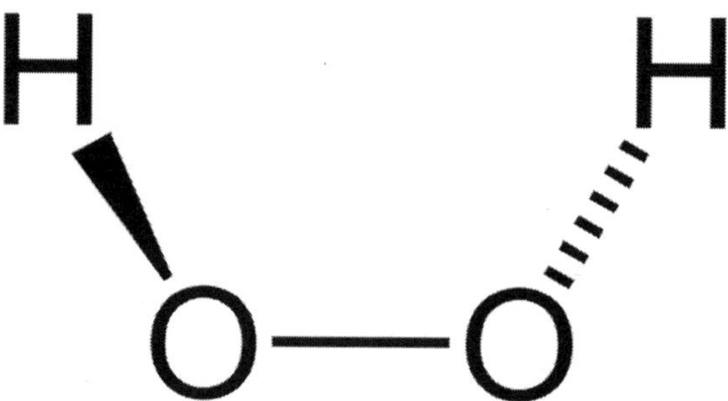

Figure: *Hydrogen peroxide*

Catalase is used in the food industry for removing hydrogen peroxide from milk prior to cheese production. Another use is in food wrappers where it prevents food from oxidizing. Catalase is also used in the textile industry, removing hydrogen peroxide from fabrics to make sure the material is peroxide-free.

A minor use is in contact lens hygiene - a few lens-cleaning products disinfect the lens using a hydrogen peroxide solution; a solution containing catalase is then used to decompose the hydrogen peroxide before the lens is used again. Recently, catalase has also begun to be used in the aesthetics industry. Several mask treatments combine the enzyme with hydrogen peroxide on the face with the intent of increasing cellular oxygenation in the upper layers of the epidermis.

Catalase Test

The catalase test is also one of the main three tests used by microbiologists to identify species of bacteria. The presence of catalase enzyme in the test isolate is detected using hydrogen peroxide. If the bacteria possess catalase (i.e., are catalase-positive), when a small amount of bacterial isolate is added to hydrogen peroxide, bubbles of oxygen are observed.

The catalase test is done by placing a drop of hydrogen peroxide on a microscope slide. Using an applicator stick, a scientist touches the colony, and then smears a sample into the hydrogen peroxide drop.

- If the mixture produces bubbles or froth, the organism is said to be 'catalase-positive'. Staphylococci and Micrococci are catalase-positive. Other catalase-positive organisms include *Listeria, Corynebacterium diphtheriae, Burkholderia cepacia, Nocardia*, the family Enterobacteriaceae (*Citrobacter, E. coli, Enterobacter, Klebsiella, Shigella, Yersinia, Proteus, Salmonella, Serratia, Pseudomonas*), *Mycobacterium tuberculosis, Aspergillus*, and *Cryptococcus.*
- If not, the organism is 'catalase-negative'. *Streptococcus* and *Enterococcus* spp. are catalase-negative.

While the catalase test alone cannot identify a particular organism, combined with other tests, such as antibiotic resistance, it can aid identification. The presence of catalase in bacterial cells depends on both the growth condition and the medium used to grow the cells.

Capillary tubes may also be used. A small amount of bacteria is collected on the end of the capillary tube (it is essential to ensure that the end is not blocked, otherwise it may present a false negative). The opposite end is then dipped into hydrogen peroxide which will draw up the liquid (through capillary action), and turned upside down, so the bacterial end is closest to the bench. A few taps of the arm should then move the hydrogen peroxide closer to the bacteria. When the hydrogen peroxide and bacteria are touching, bubbles may begin to rise, giving a positive catalase result.

Gray Hair

According to recent scientific studies, low levels of catalase may play a role in the graying process of human hair. Hydrogen peroxide is naturally produced by the body and catalase breaks it down. If catalase levels decline, hydrogen peroxide cannot be broken down as well. This allows the hydrogen peroxide to bleach the hair from the inside out. This finding may someday be incorporated into cosmetic treatments for graying hair.

Interactions

Catalase has been shown to interact with the *ABL2* and *Abl* genes.

Dehydrogenase

A dehydrogenase (also called DHO in the literature) is an enzyme that oxidizes a substrate by a reduction reaction that transfers one or more hydrides (H^-) to an electron acceptor, usually NAD^+/$NADP^+$ or a flavin coenzyme such as FAD or FMN.

Examples

- aldehyde dehydrogenase
- acetaldehyde dehydrogenase
- alcohol dehydrogenase
- glutamate dehydrogenase (an enzyme that can convert glutamate to α-Ketoglutarate and vice versa).
- lactate dehydrogenase
- pyruvate dehydrogenase (a common enzyme that feeds the TCA Cycle in converting Pyruvate to Acetyl CoA)
- glucose-6-phosphate dehydrogenase (involved in the pentose phosphate pathway)
- glyceraldehyde-3-phosphate dehydrogenase (involved in glycolysis)
- sorbitol dehydrogenase

TCA cycle examples:

- isocitrate dehydrogenase
- alpha-ketoglutarate dehydrogenase
- succinate dehydrogenase
- malate dehydrogenase.

Polyphenol Oxidase

Polyphenol oxidase (PPO or monophenol monooxygenase) is a tetramer that contains four atoms of copper per molecule, and binding sites for two aromatic compounds and oxygen. The enzyme catalyses the *o*-hydroxylation of monophenols (phenol molecules in which the benzene ring contains a single hydroxyl substituent) to *o*-diphenols (phenol molecules containing two hydroxyl substituents). They can also further catalyse the oxidation of *o*-diphenols to produce *o*-quinones. It is the rapid polymerisation of *o*-quinones to produce black, brown or red pigments (polyphenols) that is the cause of fruit browning. The amino acid tyrosine contains a single phenolic ring that may be oxidised by the action of PPOs to form *o*-quinone. Hence, PPOs may also be referred to as tyrosinases.

Enzyme nomenclature differentiates between monophenol oxidase enzymes (tyrosinases) and *o*-diphenol: oxygen oxidoreductase enzymes (catechol oxidases). Therefore, please refer to the tyrosinase and catechol oxidase articles for more information on polyphenol oxidase enzymes.

A mixture of monophenol oxidase and catechol oxidase enzymes is present in nearly all plant tissues, and can also be found in bacteria, animals, and fungi. In insects, cuticular polyphenol oxidases are present and their products are responsible for desiccation tolerance.

In fact, browning by PPO is not always an undesirable reaction; the familiar brown colour of tea and cocoa is developed by PPO enzymatic browning during product processing.

Tentoxin has also been used in recent research to eliminate the polyphenol oxidase activity from seedlings of higher plants. Tropolone is a grape polyphenol oxidase inhibitor. Another inhibitor of this enzyme is potassium pyrosulphite ($K_2S_2O_5$).

Catechol Oxidase

Catechol oxidase is an enzyme that catalyses the oxidation of phenols such as catechol. Catechol oxidase is a copper-containing enzyme whose activity is similar to that of tyrosinase, a related class of copper oxidases. They are ubiquitous plant enzymes that catalyze oxidation of a broad range of ortho-diphenols to the corresponding o-quinones coupled with the reduction of oxygen to water. It is different from tyrosinase, Ec 1.14.18.1, which can catalyze both the monooxygenation of monophenols and the oxidation of catechols.

Catechol oxidase carries out the oxidation of phenols such as catechol, using dioxygen (O_2). In the presence of catechol, benzoquinone is formed. Hydrogens removed from catechol combine with oxygen to form water. This reaction, producing the yellow compound benzoquinone, is a form of enzymatic browning exhibited in many foods upon exposure to oxygen (e.g., in bananas). The benzoquinone is then oxidised by the air to form dark-brown melanin.

Catechol is present in small quantities in the vacuoles of cells of many plant tissues. Catechol oxidase is present in the cell cytoplasm. If the plant tissues are damaged, the catechol is released and the enzyme converts the catechol to ortho-quinone, which is a natural antiseptic. Catechol oxidase, therefore, has a role in plant defence mechanisms, helping to protect damaged plants against both bacterial and fungal disease.

It has been suggested that the quantity of catechol oxidase produced by a plant may be related to its susceptibility to fungal infection. Benzoquinone inhibits the growth of microorganisms and prevents damaged fruit from rotting.

Catechol oxidase activity is also of economic importance. It has been estimated that half the world's fruit and vegetable crop is lost due to post-harvest browning reactions due to the enzyme.

Catechol + $1/2O_2$ —(Catechol Oxidase)→ Benzoquinone + H_2O

Catechol oxidase has been studied from sweet potatoes and assigned into a biological assembly. The catalytic copper centre is accommodated in a central four-helix-bundle located in a hydrophobic pocket close to the surface. Both metal binding sites are composed of three histidine ligands. His 109, ligated to the CuA site, is covalently linked to Cys 92 by an unusual thioether bond. An additional ligand component is Cu2O and contains a Cu-O-Cu linkage.

Peroxidase

Peroxidases (EC number 1.11.1.x) are a large family of enzymes that typically catalyze a reaction of the form:

$$\text{ROOR'} + \text{electron donor } (2\ e^-) + 2H^+ \rightarrow \text{ROH} + \text{R'OH}$$

For many of these enzymes the optimal substrate is hydrogen peroxide, but others are more active with organic hydroperoxides such as lipid peroxides. Peroxidases can contain a heme cofactor in their active sites, or alternately redox-active cysteine or selenocysteine residues.

The nature of the electron donor is very dependent on the structure of the enzyme.

- For example, horseradish peroxidase can use a variety of organic compounds as electron donors and acceptors. Horseradish peroxidase has an accessible active site, and many compounds can reach the site of the reaction.
- Because there is a very closed active site, for an enzyme such as cytochrome c peroxidase, the compounds that donate electrons are very specific.

While the exact mechanisms have yet to be elucidated, peroxidases are known to play a part in increasing a plant's defences against pathogens. Peroxidases are sometimes used as histological marker. Cytochrome c peroxidase is used as a soluble, easily purified model for cytochrome c oxidase.

The glutathione peroxidase family consists of 8 known human isoforms. Glutathione peroxidases use glutathione as an electron donor and are active with both hydrogen peroxide and organic hydroperoxide substrates. Gpx1, Gpx2, Gpx3, and Gpx4 have been shown to be selenium-containing enzymes, whereas Gpx6 is a selenoprotein in humans with cysteine-containing homologues in rodents.

Amyloid beta, when bound to heme, has been shown to have peroxidase activity.

A typical group of peroxidases are the haloperoxidases. This group is able to form reactive halogen species and, as a result, natural organohalogen substances.

A majority of peroxidase protein sequences can be found in the PeroxiBase database.

Applications

Peroxidase can be used for treatment of industrial waste waters. For example, phenols, which are important pollutants, can be removed by enzyme-catalyzed polymerization using horseradish peroxidase. Thus phenols are oxidized to phenoxy radicals, which participate in reactions where polymers and oligomers are produced that are less toxic than phenols.

Furthermore, peroxidases can be made out of the fecal matter of rats and chickens and eaten off human skin. There are many investigations about the use of peroxidase in many manufacturing processes like adhesives, computer chips, car parts, and linings of drums and cans. Other studies have shown that peroxidases may be used successfully to polymerize anilines and phenols in organic solvent matrices.

Proteolysis

Proteolysis is the breakdown of proteins into smaller polypeptides or amino acids. In general, this occurs by the hydrolysis of the peptide bond, and is most commonly achieved by cellular enzymes called proteases, but may also occur by intramolecular digestion, as well as by non-enzymatic methods such as the action of mineral acids and heat.

Proteolysis in organisms serves many purposes; for example, digestive enzymes break down proteins in food to provide amino acids for the organism, while proteolytic processing of polypeptide chain after its synthesis may be necessary for the production of an active protein. It is also important in the regulation of some physiological and cellular processes, as well as preventing the accumulation of unwanted or abnormal proteins in cells.

Post-translational Proteolytic Processing

Limited proteolysis of a polypeptide during or after translation in protein synthesis often occur for many proteins. This may involve removal of the N-terminal methionine, signal peptide, and/or the conversion of an inactive or non-functional protein to an active one. The precursor to the final functional form of protein is termed proprotein, and these proproteins may be first synthesized as preproprotein. For example, albumin is first synthesized as preproalbumin and contains an uncleaved signal peptide. This forms the proalbumin after the signal peptide is cleaved, and a further processing to remove the N-terminal 6-residue propeptide yields the mature form of the protein.

Removal of N-terminal Methionine

The initiating methonine (and, in prokaryotes, fMet) may be removed during translation of the nascent protein. For *E. coli*, fMet is efficiently removed if the second residue is small and uncharged, but not if the second residue is bulky and charged. In both prokaryotes and eukaryotes, the exposed N-terminal residue may determine the half-life of the protein according to the N-end rule.

Removal of the Signal Sequence

Proteins that are to be targeted to a particular organelle or for secretion have an N-terminal signal peptide that directs the protein to its final destination. This signal peptide is removed by proteolysis after their transport through a membrane.

Cleavage of Polyprotein

Some proteins and most eukaryotic polypeptide hormones are synthesized as a large precursor polypeptide known as polyprotein that require proteolytic cleavage into individual smaller polypeptide chains. The polyprotein pro-opiomelanocortin (POMC) contains many polypeptide hormones. The cleavage pattern of POMC, however, may vary between different tissues, yielding different sets of polypeptide hormones from the same polyprotein.

Many viruses also produce their proteins initially as a single polypeptide chain that were translated from a polycistronic mRNA. This polypeptide is subsequently cleaved into individual polypeptide chains.

Cleavage of Precursor Proteins

Many proteins and hormones are synthesized in the form of their precursors - zymogens, proenzymes, and prehormones. These proteins are cleaved to form their final active structures. Insulin, for example, is synthesized as preproinsulin, which yields proinsulin after the signal peptide has been cleaved. To form the mature insulin, the proinsulin is then cleaved at two positions to yield two polypeptide chains linked by 2 disulphide bonds. Proinsulin is necessary for the folding of the polypeptide chain, as the 2 polypeptide chains of insulin may not correctly assemble into the correct form, whereas its precursor proinsulin does.

Proteases in particular are synthesized in the inactive form so that they may be safely stored in cells, and ready for release in sufficient quantity when required. This is to ensure that the protease is activated only in the correct location or context, as inappropriate activation of these proteases can be very destructive for an organism. Proteolysis of the zymogen yields an active protein; for example, when trypsinogen is cleaved to form trypsin, a slight rearrangement of the protein structure that completes the active site of the protease occurs, thereby activating the protein.

Proteolysis can, therefore, be a method of regulating biological processes by turning inactive proteins into active ones. A good example is the blood clotting cascade whereby an initial event triggers a cascade of sequential proteolytic activation of many specific proteases, resulting in blood coagulation. The complement system of the immune response also involves a complex sequential proteolytic activation and interaction that result in an attack on invading pathogens.

Protein Degradation

Proteolytic cleavage breaks down proteins in food extracellularly into smaller peptides and amino acids so that they may be absorbed and used by an organism. Proteins in cells are also constantly being broken down into amino acids. This intracellular degradation of protein serves a number of functions: It removes damaged and abnormal protein and prevent their accumulation, and it also serves to regulate cellular processes by removing enzymes and regulatory proteins that are no longer needed. The amino acids may then be reused for protein synthesis.

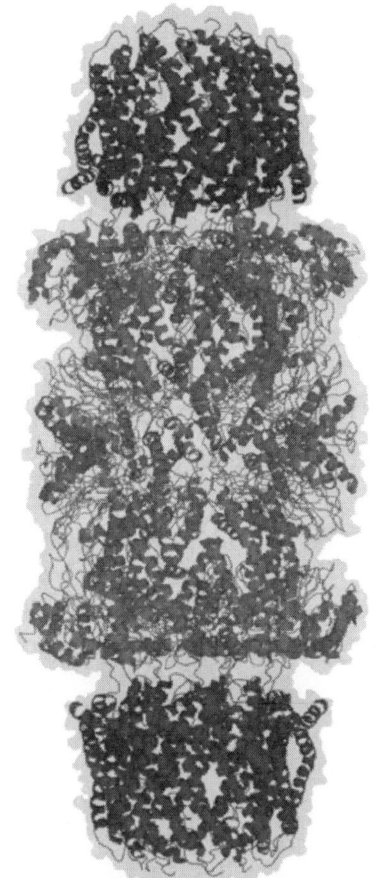

Figure: *Structure of a proteasome. Its active sites are inside the tube (blue) where proteins are degraded.*

Lysosome and Proteasome

The intracellular degradation of protein may be achieved in two ways - proteolysis in lysosome, or a ubiquitin-dependent process that targets unwanted proteins to proteasome. The autophagy-lysosomal pathway is normally a non-selective process, but it may become selective upon starvation whereby proteins with peptide sequence KFERQ or similar are selectively broken down. The lysosome contains a large number of proteases such as cathepsins.

The ubiquitin-mediated process is selective. Proteins marked for degradation are covalently linked to ubiquitin. Many molecules of ubiquitin may be linked in tandem to a protein destined for degradation. The polyubiquinated protein is targeted to an ATP-dependent protease complex, the proteasome. The ubiquitin is released and reused, while the targeted protein is degraded.

Rate of Intracellular Protein Degradation

Different proteins are degraded at different rate. Abnormal proteins are quickly degraded, whereas the rate of degradation of normal proteins may vary widely depending on their functions. Enzymes at important metabolic control points may be degraded much faster than those enzymes whose activity is largely constant under all physiological conditions. One of the most rapidly degraded proteins is ornithine decarboxylase, which has a half-life of 11 minutes. In contrast, other proteins like actin and myosin have half-life of a month or more, while, in essence, haemoglobin lasts for the entire life-time of erythrocyte.

The N-end rule may partially determine the half-life of a protein, and proteins with segments rich in proline, glutamine, serine, and threonine (the so-called PEST proteins) have short half-life. Other factors suspected to affect degradation rate include the rate deamination of glutamine and asparagine and oxidation of cystein, histidine, and methionine, the absence of stabilizing ligands, the presence of attached carbohydrate or phosphate groups, the presence of free α-amino group, the negative charge of protein, and the flexibility and stability of the protein.

The rate of proteolysis may also depend on the physiological state of the cell, such as its hormonal state as well as nutritional status. In time of starvation, the rate of protein degradation increases.

Digestion

In human digestion, proteins in food are broken down into smaller peptide chains by digestive enzymes such as pepsin, trypsin, chymotrypsin, and elastase, and into amino acids by various enzymes such as carboxypeptidase, aminopeptidase, and dipeptidase. It is necessary to break down proteins into small peptides (tripeptides and dipeptides) and amino acids so they can be absorbed by the intestines, and the absorbed tripeptides and dipeptides are also further broken into amino acids intracellularly before they enter the bloodstream. Different enzymes have different specificity for their substrate; trypsin, for example, cleaves the peptide bond after a positively charged residue (arginine and lysine); chymotrypsin cleaves the bond after an aromatic residue (phenylalanine, tyrosine, and tryptophan); elastase cleaves the bond after a small non-polar residue such as alanine or glycine.

In order to prevent inappropriate or premature activation of the digestive enzymes (they may, for example, trigger pancreatic self-digestion causing pancreatitis), these enzymes are secreted as inactive zymogen. The precursor of pepsin, pepsinogen, is secreted by the stomach, and is activated only in the acidic environment found in stomach. The pancreas secretes the precursors of a number of proteases, such as trypsin and chymotrypsin. The zymogen of trypsin is trypsinogen, which is activated by a very specific protease, enterokinase, secreted by the mucosa of the duodenum. The trypsin, once activated, can also cleave other trypsinogens as well as the precursors of other proteases such as chymotrypsin and carboxypeptidase.

In bacteria, a similar strategy of employing an inactive zymogen or prezymogen is used. Subtilisin, which is produced by *Bacillus subtilis*, is produced as preprosubtilisin, and is released only if the signal peptide is cleaved and autocatalytic proteolytic activation has occurred.

Proteolysis in Cellular Regulation

Proteolysis is also involved in the regulation of many cellular processes by activating or deactivating enzymes, transcription factors, and receptors, for example in the biosynthesis of cholesterol, or the mediation of thrombin signalling through protease-activated receptors.

Some enzymes at important metabolic control points such as ornithine decarboxylase is regulated entirely by its rate of synthesis and its rate of degradation. Other rapidly degraded proteins include the protein products of proto-oncogenes, which play central roles in the regulation of cell growth.

Cell Cycle Regulation

Cyclins are a group of proteins that activate kinases involved in cell division. The degradation of cyclins is the key step that governs the exit from mitosis and progress into the next cell cycle. Cyclins accumulate in the course the cell cycle, then abruptly disappear just before the anaphase of mitosis. The cyclins are removed via a ubiquitin-mediated proteolytic pathway.

Apoptosis

Caspases are an important group of proteases involved in apoptosis. The precursors of caspase, procaspase, may be activated by proteolysis through its association with a protein complex that forms apoptosome, or by granzyme B, or via the death receptor pathways.

Regulatory Domains in Proteolysis

Protease may have one or more regulatory domains -

- Calcium-binding domain - e.g., prothrombin, factor IX, X, VII, protein C in blood clotting cascade, calpain.
- Kringle domain - e.g., in prothrombin it keeps the protease inactive.

Proteolysis and Diseases

Abnormal proteolytic activity are associated with many diseases. In pancreatitis, leakage of proteases and their premature activation in the pancreas results in the self-digestion of the pancreas. People with diabetes mellitus may have increased lysosomal activity and the degradation of some proteins can increase significantly. Chronic inflammatory diseases such as rheumatoid arthritis may involve the release of lysosomal enzymes into extracellular space that break down surrounding tissues. Abnormal proteolysis and generation of peptides that aggregate in cells and their ineffective removal may result in many age-related neurological diseases such as Alzheimer's.

Other diseases linked to aberrant proteolysis include muscular dystrophy, degenerative skin disorders, respiratory and gastrointestinal diseases, and malignancy.

Non-enzymatic Proteolysis

Chemicals may be used in laboratory to target specific residue and cleave its peptide bond so that protein may be broken down into smaller polypeptides for analysis. Cyanogen bromide is often used to cleave the peptide bond after a methionine. Other methods may be used to specifically cleave tryptophanyl, aspartyl, cysteinyl, and asparaginyl peptide bonds. Acids such as trifluoroacetic acid and formic acid may also be used.

Strong mineral acids can readily hydrolyse the peptide bonds in a protein. However, some proteins are remarkably resistant to hydrolysis. One well-known example is ribonuclease A, and one method for its purification involves treatment of crude extracts with hot sulphuric acid so that other proteins become degraded while ribonuclease A is left intact.

Laboratory Applications

Proteolysis is also used in research and diagnostic applications:

- Cleavage of fusion protein so that the fusion partner and protein tag used in protein expression and purification may be removed. The proteases used have high degree of specificity, such as thrombin, enterokinase, and TEV protease, so that only the targeted sequence may be cleaved.
- Complete inactivation of undesirable enzymatic activity or removal of unwanted proteins. For example, proteinase K, a broad-spectrum proteinase stable in urea and SDS, is often used in the preparation of nucleic acids to remove unwanted nuclease contaminants that may otherwise degrade the DNA or RNA.
- Partial inactivation, or changing the functionality, of specific protein. For example, treatment of DNA polymerase I with subtilisin yields the Klenow fragment, which retains its polymerase function but lacks 5'-exonuclease activity.
- In-gel digestion of proteins after separation by gel electrophoresis for the identification by mass spectrometry.
- Digestion of proteins in solution for proteome analysis by liquid chromatography-mass spectrometry (LC-MS).
- Analysis of the stability of folded domain under a wide range of conditions.
- Increasing success rate of crystallisation projects

Venoms

Certain types of venom, such as those produced by venomous snakes, can also cause proteolysis. These venoms are, in fact, complex digestive fluids that begin their work outside of the body. Proteolytic venoms cause a wide range of toxic effects, including effects that are:

- cytotoxic (cell-destroying)
- hemotoxic (blood-destroying)
- myotoxic (muscle-destroying)
- hemorrhagic (bleeding)

Urease

Ureases (EC 3.5.1.5), functionally, belong to the superfamily of amidohydrolases and phosphotriestreases. It is an enzyme that catalyzes the hydrolysis of urea into carbon dioxide and ammonia. The reaction occurs as follows:

$$(NH_2)_2CO + H_2O \rightarrow CO_2 + 2NH_3$$

More specifically, urease catalyzes the hydrolysis of urea to produce ammonia and carbamate, the carbamate produced is subsequently degraded by spontaneous hydrolysis to produce another ammonia and carbonic acid. Urease activity tends to increase the pH of the environment in which it is as it produces ammonia, as it is a basic molecule. Ureases are found in numerous bacteria, fungi, algae, plants and some invertebrates, as well as in soils, as a soil enzyme. They are nickel-containing metalloenzymes of high molecular weight.

In 1926, James B. Sumner, an assistant professor at Cornell University, showed that urease is a protein by examining its crystallized form. Sumner's work was the first demonstration that a pure protein can function as an enzyme, and led eventually to the recognition that most enzymes are in fact proteins, and the award of the Nobel prize in chemistry to Sumner in 1946. The structure of urease was first solved by P. A. Karplus in 1995. Urease was the first ever enzyme crystallized.

Characteristics

- Active site requiring nickel in Jack Beans and several bacteria. However, in vitro activation also has been achieved with manganese and cobalt
- Molecular weight: 480 kDa or 545 kDa for Jack Bean Urease (calculated mass from the amino acid sequence). 840 amino acids per molecule, of which 90 are cysteines.
- Optimum pH: 7.4
- Optimum Temperature: 60 degrees Celsius
- Enzymatic specificity: urea and hydroxyurea
- Inhibitors: heavy metals (Pb^- & Pb^{2+})

Bacterial ureases are composed of three distinct subunits, one large (α 60–76kDa) and two small (β 8–21 kDa, γ 6–14 kDa) commonly forming (αβγ)3 trimers stoichiometry with a 2-fold symmetric structure (note that the image above gives the structure of the asymmetric unit, one-third of the true biological assembly), they are cysteine-rich enzymes, resulting in the enzyme molar masses between 190 and 300kDa.

An exceptional enzyme is the urease of Helicobacter species, which is composed of two subunits, α(61–66 kDa)-β(26–31 kDa), and has been shown to form a supramolecular dodecameric complex. of repeating α-β subunits, each coupled pair of subunits has an active site, for a total of 12 active sites. ($\alpha_{12}\beta_{12}$). It plays an essential function for survival, neutralizing gastric acid by allowing urea to enter into periplasm via a proton-gated urea channel. The presence of urease is used in the diagnosis of *Helicobacter* species.

All bacterial ureases are solely cytoplasmic, except for Helicobacter pylori urease, which along with its cytoplasmic activity, has external activity with host cells. In contrast, all plant ureases are cytoplasmic.

Fungal and plant ureases are made up of identical subunits (~90 kDa each), most commonly assembled as trimers and hexamers. For example, Jack Bean urease has two structural and one catalytic subunit. The α subunit contains the active site, it is composed of 840 amino acids per molecule (90 cysteines), its molecular mass without Ni(II) ions amounting to 90.77 kDa. The mass of the hexamer with the 12 nickel ions is 545.34 kDa. It is structurally related to the (αβγ)3 trimer of bacterial ureases. Other examples of homohexameric structures of plant ureases are those of soybean, pigeon pea and cotton seeds enzymes.

It is important to note, that although composed of different types of subunits, ureases from different sources extending from bacteria to plants and fungi exhibit high homology of amino acid sequences.

Active Site

The active site of all ureases known are located in the α (alpha) subunits. It is a bis-μ-hydroxo dimeric nickel centre, with an interatomic distance of ~3.5 Å, Magnetic susceptibility experiments have indicated that, in Jack Bean Urease, high spin octahedrally coordinated Ni(II) ions are weakly antiferromagnetically coupled. The Nickel ions bridged by a carbamylated lysine through its O-atoms and by a hydroxide ion. Ni(1) is coordinated by N-atoms of histidines residues and one water molecule, is it said to be pseudo square pyramidal. Ni(2) is coordinated by two histidines also through N-atoms and additionally by aspartic acid through its O- atom and 2 water molecules. X-ray absorption spectroscopy (XAS) studies of Canavalia ensiformis (jack bean), Klebsiella aerogenes and Sporosarcina pasteurii (formerly known as

"Bacillus pasteurii") confirm 5–6 coordinate nickel ions with exclusively O/N ligands (two imidazoles per nickel).

The water molecules are located towards the opening of the active site and form a tetrahedral cluster that fills the cavity site through Hydrogen bonds, and it's here where urea binds to the active site for the reaction, displacing the water molecules. The amino acid residues participate in the substrate binding, mainly through H-bonding, stabilize the catalytic transition state and accelerate the reaction. Additionally, the amino acid residues involved in the architecture of the active site compose part of the mobile flap of the site, which is said to act as a gate for the substrate. Cysteine residues are common in the flap region of the enzymes, which have been determined not to be essential in catalysis, although involved in positioning other key residues in the active site appropriately. In the structure of Sporosarcina pasteurii urease the flap was found in the open conformation, while its closed conformation is apparently needed for the reaction.

When compared, the α subunits of Helicobacter pylori urease and other bacterial ureases align with the jack bean urease's, suggesting that all ureases are evolutionary variants of one ancestral enzyme.

Note: It is important to note that the coordination of urea to the active site of urease has never been observed in a resting state of the enzyme.

Urease Activity

The kcat/Km of urease in the processing of urea is 10^{14} times greater than the rate of the uncatalyzed elimination reaction of urea. There are many reasons for this observation in nature. The proximity of urea to active groups in the active site along with the correct orientation of urea allow hydrolysis to occur rapidly. Urea alone is very stable due to the resonance forms it can adopt. The stability of urea is understood to be due to its resonance energy, which has been estimated at 30-40 kcal/mol. This is because the zwitterionic resonance forms all donate electrons to the carbonyl carbon making it less of an electrophile making it less reactive to nucleophilic attack.

Proposed Mechanisms of Urease

The Blakeley/Zerner Proposed Mechanism: One mechanism for the catalysis of this reaction by urease was proposed by Blakely and Zerner. It begins with a nucleophilic attack by the carbonyl oxygen of the urea molecule onto the 5-coordinate Ni (Ni-1). A weakly coordinated water ligand is displaced in its place. A lone pair of electrons from one of the nitrogen atoms on the Urea molecule creates a double bond with the central carbon, and the resulting NH_2^+ of the coordinated substrate interacts with a nearby negatively charged group. Blakeley and Zerner proposed this nearby group to be a Carboxylate ion

A hydroxide ligand on the six coordinate Ni is deprotonated by a base. The carbonyl carbon is subsequently attacked by the electronegative oxygen. A pair of electrons from the nitrogen-carbon double bond returns to the nitrogen and neutralizes the charge on it, while the now 4-coordinate carbon assumes an intermediate tetrahedral orientation.

The breakdown of this intermediate is then helped by a sulfhydryl group of a cysteine located near the active site. A hydrogen bonds to one of the nitrogen atoms, breaking its bond with carbon, and releasing an NH3 molecule. Simultaneously, the bond between the oxygen and the 6-coordinate nickel is broken. This leaves a carbamate ion coordinated to the 5-coordinate Ni, which is then displaced by a water molecule, regenerating the enzyme.

The carbamate produced then sponaneously degrades to produce another ammonia and carbonic acid.

The Hausinger/Karplus Proposed Mechanism

The mechanism proposed by Hausinger and Karplus attempts to revise some of the issues apparent in the Blakely and Zerner pathway, and focuses on the positions of the side chains making up the urea-binding pocket. From the crystal structures from K. aerogenes urease, it was argued that the general base used in the Blakely mechanism, His^{320}, was too far away from the Ni2-bound water to deprotonate in order to form the attacking hydroxide moiety. In addition, the general acidic ligand required to protonate the urea nitrogen was not identified. Hausinger and Karplus suggests a reverse protonation scheme, where a protonated form of the His^{320} ligand plays the role of the general acid and the Ni2-bound water is already in the deprotonated state. The mechanism follows the same path, with the general base omitted (as there is no more need for it) and His^{320} donating its proton to form the ammonia molecule, which is then released from the enzyme. While the majority of the His^{320} ligands and bound water will not be in their active forms (protonated and deprotonated, respectively,) it was calculated that approximately 0.3% of total urease enzyme would be active at any one time. While logically, this would imply that the enzyme is not very efficient, contrary to established knowledge, usage of the reverse protonation scheme provides an advantage in increased reactivity for the active form, balancing out the disadvantage. Placing the His^{320} ligand as an essential component in the mechanism also takes into account the mobile flap region of the enzyme. As this histidine ligand is part of the mobile flap, binding of the urea substrate for catalysis closes this flap over the active site and with the addition of the hydrogen bonding pattern to urea from other ligands in the pocket, speaks to the selectivity of the urease enzyme for urea.

The Ciurli/Mangani Proposed Mechanism

The mechanism proposed by Ciurli and Mangani is one of the more recent and currently accepted views of the mechanism of urease and is based primarily on the different roles of the two nickel ions in the active site. One of which binds and activates urea, the other nickel ion binds and activates the nucleophilic water molecule. With regards to this proposal, urea enters the active site cavity when the mobile 'flap' (which allows for the entrance of urea into the active site) is open. Stability of the binding of urea to the active site is achieved via a hydrogen-bonding network, orienting the substrate into the catalytic cavity. Urea binds to the five-coordinated nickel (Ni1) with the carbonyl oxygen atom. It approaches the six-coordinated nickel (Ni2) with one of its amino groups and subsequently bridges the two nickel centres. The binding of the urea carbonyl oxygen atom to Ni1 is stabilized through the protonation state of $His^{\alpha 222}$ N. Additionally, the conformational change from the open to closed state of the mobile flap generates a rearrangement of $Ala^{\alpha 222}$ carbonyl group in such a way that its oxygen atom points to Ni2. The $Ala^{\alpha 170}$ and $Ala^{\alpha 366}$ are now oriented in a way that their carbonyl groups act as hydrogen-bond acceptors towards NH_2 group of urea, thus aiding its binding to Ni2. Urea is a very poor chelating ligand due to low Lewis base character of its NH_2 groups. However the carbonyl oxygens of $Ala^{\alpha 170}$ and $Ala^{\alpha 366}$ enhance the basicity of the NH_2 groups and allow for binding to Ni2. Therefore in this proposed mechanism, the positioning of urea in the active site is induced by the structural features of the active site residues which are positioned to act as hydrogen-bond donors in the vicinity of Ni1 and as acceptors in the vicinity of Ni2. The main structural difference between the Ciurli/Mangani mechanism and the other two are that it incorporates a nitrogen, oxygen bridging urea that is attacked by a bridging hydroxide.

Action of Urease in Pathogenesis

Bacterial ureases are most often the mode of pathogenesis for many medical conditions. They are associated with hepatic encephalopathy / Hepatic coma, infection stones, and peptic ulceration.

Infection Stones

Infection induced urinary stones are a mixture of struvite ($MgNH_4PO_4 \cdot 6H_2O$) and carbonate apatite [$Ca_{10}(PO_4)6 \cdot CO_3$]. These polyvalent ions are soluble but become insoluble when ammonia is produced from microbial urease during urea hydrolysis, as this increases the surrounding environments pH from roughly 6.5 to 9. The resultant alkalinization results in stone crystallization. In humans the microbial urease, *Proteus mirabilis*, is the most common in infection induced urinary stones.

Urease in Hepatic Encephalopathy/ Hepatic coma

Studies have shown that *Helicobacter pylori* along with cirrhosis of the liver cause hepatic encephalopathy and hepatic coma. *Heliobacter pylori* are microbial ureases found in the stomach. As ureases they hydrolyze urea to produce ammonia and carbonic acid. As the bacteria are localized to the stomach ammonia produced is readily taken up by the circulatory system from the gastric lumen. This results in elevated ammonia levels in the blood and is coined as hyperammonemia, eradication of *Heliobacter pylori* show marked decreases in ammonia levels.

Urease in Peptic Ulcers

Helicobacter pylori is also the cause of peptic ulcers with its manifestation in 55%-68% reported cases. . This was confirmed by decreased ulcer bleeding and ulcer reoccurrence after eradication of the pathogen. In the stomach there is an increase in pH of the mucosal lining as a result of urea hydrolysis this prevents movement of hydrogen ions between gastric glands and gastric lumen. In addition, the high ammonia concentrations have an effect on intercellular tight junctions increasing permeability and also disrupting the gastric mucous membrane of the stomach.

As Diagnostic Test

Many gastrointestinal or urinary tract pathogens produce urease, enabling the detection of urease to be used as a diagnostic to detect presence of pathogens.

Urease-positive pathogens include:

- *Proteus mirabilis* and *Proteus vulgaris*
- *Ureaplasma urealyticum*, a relative of *Mycoplasma* spp.
- *Nocardia*
- *Campylobacter ureolyticus*
- *Cryptococcus* spp., an opportunistic fungus
- *Helicobacter pylori*
- Certain Enteric bacteria including *Proteus* spp., *Klebsiella* spp., *Morganella, Providencia,* and possibly *Serratia* spp.
- *Brucella*

Other Uses

Urease conductometric biosensors for detection of heavy-metal ions. Urease conductometric biosensors for detection of heavy-metal ions consisting of interdigitated gold electrodes and enzyme membranes formed on their sensitive parts have been used for a quantitative estimation of general water pollution with heavy-metal ions. The measurements of the urease residual activity have been carried out in Tris-$HNO_{3>}$ buffer after preincubation in model metal-salt solution. The detection limits, depending on preincubation time and dynamic ranges, have been determined in model solutions of heavy-metal ions. The sequence of metals ions relative to their toxicity toward urease is: Hg^{2+} > Cu^{2+} > Cd^{2+} > Co^{2+} > Pb^{2+} > Sr^{2+} > . The conditions for practical applications of the biosensors have been investigated and critically evaluated for optimization. Urease reactivation by EDTA after inhibition by heavy-metal ions has been demonstrated. The performance characteristics of the conductometric biosensor are discussed by G. A. Zhylyaka, S. V. Dzyadevichb, Y. I. Korpana, A. P. Soldatkina and A. V. El'skayaa in their paper.

Soil Texture

Soil texture is a qualitative classification tool used in both the field and laboratory to determine classes for agricultural soils based on their physical texture. The classes are distinguished in the field by the 'textural feel' which can be further clarified by separating the relative proportions of sand, silt and clay using grading sieves: The Particle-size distribution (PSD). The class is then used to determine crop suitability and to approximate the soils responses to environmental and management conditions such as drought or calcium (lime) requirements. A qualitative rather than a quantitative tool it is a fast, simple and effective means to assess the soils physical characteristics.

Although the U.S.D.A. system uses 12 classes whilst the U.K.-ADAS uses just 11, the systems are mutually compatible as shown in the combined soil textural triangle below.

Field Assessment Using a Hand Texture Chart

Hand analysis, whilst an arbitrary technique, is an extremely simple and effective means to rapidly assess and classify a soil's physical condition. Correctly executed the procedure allows for rapid and frequent assessment of soil characteristics with little or no equipment. It is thus an extremely useful tool for identifying spatial variation both within and between plots (fields) as well as identifying progressive changes and boundaries between soil classes and orders.

The method involves taking a small sample of soil, sufficient to roll into a ball of approx 2.5 cm diameter, from just below the surface. Using a small drop of water or 'spit' the sample is then moisten to the sticky point (the point at which it begins to adhere to the finger). The ball is then molded to determine its workability and its class according to the steps in the chart opposite.

Soil Separates

Soil separates are specific ranges of particle sizes. In the United States, the smallest particles are *clay* particles and are classified by the USDA as having diameters of less than 0.002 mm. The next smallest particles are *silt* particles and have diameters between 0.002 mm and 0.05 mm. The largest particles are *sand* particles and are larger than 0.05 mm in diameter. Furthermore, large sand particles can be described as *coarse*, intermediate as *medium*, and the smaller as *fine*. Other countries have their own particle size classifications.

Name of soil separate	***Diameter limits (mm) (USDA classification)***
Clay	less than 0.002
Silt	0.002–0.05
Very fine sand	0.05–0.10
Fine sand	0.10–0.25
Medium sand	0.25–0.50
Coarse sand	0.50–1.00
Very coarse sand	1.00–2.00

Soil Texture Classification

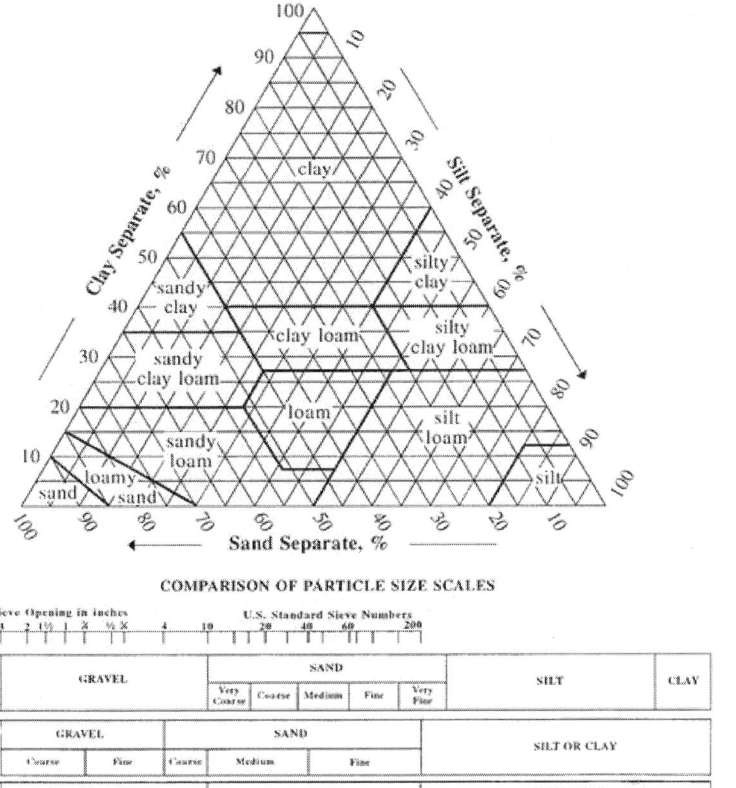

Figure: *Soil texture triangle, showing the 12 major textural classes, and particle size scales as defined by the USDA.*

Soil textures are classified by the fractions of each soil separate (sand, silt, and clay) present in a soil. Classifications are typically named for the primary constituent particle size or a combination of the most abundant particles sizes, e.g. "sandy clay" or "silty clay." A fourth term, loam, is used to describe a roughly equal concentration of sand, silt, and clay, and lends to the naming of even more classifications, e.g. "clay loam" or "silt loam."

In the United States, twelve major soil texture classifications are defined by the USDA.

Determining the soil textures is often aided with the use of a soil texture triangle.'

History of Classification

The first classification, the International system, was first proposed by Albert Atterberg (1905), and was based on his studies in southern Sweden. Atterberg chose 20 µm for the upper limit of silt fraction because particles smaller than that size were not visible to the naked eye, the suspension could be coagulated by salts, capillary rise within 24 hours was most rapid in this fraction, and the pores between compacted particles were so small as to prevent the entry of root hairs. Commission One of the International Society of Soil Science (ISSS) recommended its use at the first International Congress of Soil Science in Washington in 1927. Australia adopted this system and according to Marshall (1947) its equal logarithmic intervals are an attractive feature worth maintaining. The USDA adopted its own system in 1938, and the FAO used the USDA system in the FAO-UNESCO world soil map and recommended its use.

5

Microfauna and Microflora

Microfauna are the smallest of the soil fauna and are less than 0.1 mm in size, and so need a microscope to be seen. The two most important soil creatures are the nematodes and the protozoa. Nematodes occur widely in soils, particularly in sandy soils. They depend on a thin film of water around particles for their movement. Some species are parasites and can be a problem for agricultural crops such as potatoes. Protozoa are small and variable in shape. They are major consumers of bacteria. They are well suited to life in soil because they slide over surfaces relatively easily, feeding on soil particles, roots and thin water films in the soil.

There are three main forms of microflora in soils: bacteria, fungi and viruses. Bacteria are tiny organisms composed of single cells and without a distinct nucleus. They are extremely numerous in soils with billions in just one gram of soil and many thousands of species also within a single gram. Bacteria take part in some of the most important transformations in soils including weathering of rocks and minerals, breakdown of organic matter, and many aspects of nutrient cycling. Fungi are also very common in soils, taking the form of spores, globules and filaments. They depend on living and dead matter in the soil for their carbon and energy.

They are important in the decomposition of organic matter and also play an important part in stabilising soil aggregates. Very importantly, mycorrhizal fungi play a major part in securing nutrients for plant production and many plants are dependent on such relationships.

Viruses are the smallest and simplest multiplying entities in the soil but, perhaps because of their small size, rather little is known about them. All are parasites, i.e. they live off other flora and fauna. A range of plant, insect and human viruses can be found in soils. The conditions in soils that most influence the numbers of viruses are moisture, the surface of soil aggregates and structural units and the rooting system of the plants. We know relatively little about the viruses in soil and also about many of the other tiny organisms. The fact that many are extremely small and are out of sight below ground means that the full importance of many of these creatures may yet remain undiscovered for some time.

Plant Nutrition

Plant Nutrition is the study of the chemical elements and compounds that are necessary for plant growth, and also of their external supply and internal metabolism. In 1972, E. Epstein defined two criteria for an element to be essential for plant growth:

1. in its absence the plant is unable to complete a normal life cycle; or
2. that the element is part of some essential plant constituent or metabolite.

This is in accordance with Liebig's law of the minimum. There are 17 essential plant nutrients. Carbon and oxygen are absorbed from the air, while other nutrients including water are obtained from the soil. Plants must obtain the following mineral nutrients from the growing media:

- the primary macronutrients: nitrogen (N), phosphorus (P), potassium (K)
- the three secondary macronutrients: calcium (Ca), sulphur (S), magnesium (Mg)
- the macronutrient Silicon (Si)
- the micronutrients/trace minerals: boron (B), chlorine (Cl), manganese (Mn), iron (Fe), zinc (Zn), copper (Cu), molybdenum (Mo), nickel (Ni), selenium (Se), and sodium (Na)

The macronutrients are consumed in larger quantities and are present in plant tissue in quantities from 0.2% to 4.0% (on a dry matter weight basis). Micro nutrients are present in plant tissue in quantities measured in parts per million, ranging from 5 to 200 ppm, or less than 0.02% dry weight.

Most soil conditions across the world can provide plants with adequate nutrition and do not require fertilizer for a complete life cycle. However, man can artificially modify soil through the addition of fertilizer to promote vigorous growth and increase yield. The plants are able to obtain their required nutrients from the fertilizer added to the soil. A colloidal carbonaceous residue, known as humus, can serve as a nutrient reservoir. Besides lack of water and sunshine, nutrient deficiency is a major growth limiting factor.

Nutrient uptake in the soil is achieved by cation exchange, where root hairs pump hydrogen ions (H^+) into the soil through proton pumps. These hydrogen ions displace cations attached to negatively charged soil particles so that the cations are available for uptake by the root.

Plant nutrition is a difficult subject to understand completely, partially because of the variation between different plants and even between different species or individuals of a given clone. An element present at a low level may cause deficiency symptoms, while the same element at a higher level may cause toxicity. Further, deficiency of one element may present as symptoms of toxicity from another element. An abundance of one nutrient may cause a deficiency of another nutrient. Also a lowered availability of a given nutrient, such as SO_2^{4-} can affect the uptake of another nutrient, such as NO_3^-. Also, K^+ uptake can be influenced by the amount NH_4^+ available.

The root, especially the root hair, is the most essential organ for the uptake of nutrients. The structure and architecture of the root can alter the rate of nutrient uptake. Nutrient ions are transported to the centre of the root, the stele in order for the nutrients to reach the conducting tissues, xylem and phloem. The Casparian strip, a cell wall outside of the stele but within the root, prevents passive flow of water and nutrients to help regulate the uptake of nutrients and water. Xylem moves water and inorganic molecules within the plant and phloem counts organic molecule transportation. Water potential plays a key role in a plants nutrient uptake. If the water potential is more negative within the plant than the surrounding soils, the nutrients will move from the more higher solute (soil) concentration to lower solute concentration (plant).

There are three fundamental ways plants uptake nutrients through the root: 1.) simple diffusion, occurs when a nonpolar molecule, such as O_2, CO_2, and NH_3 that follow a concentration gradient, can passively move through the lipid bilayer membrane without the use of transport proteins. 2.) facilitated diffusion, is the rapid movement of solutes or ions following a concentration gradient, facilitated by transport proteins. 3.) Active transport, is the active transport of ions or molecules against a concentration gradient that requires an energy source, usually ATP, to pump the ions or molecules through the membrane.

- Nutrients are moved inside a plant to where they are most needed. For example, a plant will try to supply more nutrients to its younger leaves than its older ones. So when nutrients are mobile, the lack of nutrients is first visible on older leaves. However, not all nutrients are equally mobile. When a less mobile nutrient is lacking, the younger leaves suffer because the nutrient does not move up to them but stays lower in the older leaves. Nitrogen, phosphorus, and potassium are mobile nutrients, while the others have varying degrees of mobility. This phenomenon is helpful in determining what nutrients a plant may be lacking.

A symbiotic relationship may exist with 1.) Nitrogen-fixing bacteria, such as rhizobia, which are involved with nitrogen fixation, and 2.) mycorrhiza, which help to create a larger root surface area. Both of these mutualistic relationships enhance nutrient uptake.

Though nitrogen is plentiful in the Earth's atmosphere, relatively few plants engage in nitrogen fixation (conversion of atmospheric nitrogen to a biologically useful form). Most plants therefore require nitrogen compounds to be present in the soil in which they grow. These can either be supplied by decaying matter, nitrogen fixing bacteria, animal waste, or through the agricultural application of purpose made fertilizers.

Hydroponics, is growing plants in a water-nutrient solution without the use of nutrient-rich soil. It allows researchers and home gardeners to grow their plants in a controlled environment. The most common solution, is the Hoagland solution, developed by D. R. Hoagland in 1933, the solution consists of all the essential nutrients in the correct proportions necessary for most plant growth. An aerator is used to prevent an anoxic event or hypoxia. Hypoxia can affect nutrient uptake of a plant because without oxygen present, respiration becomes inhibited within the root cells. The Nutrient film technique is a variation of hydroponic technique. The roots are not fully submerged, which allows for adequate aeration of the roots, while a "film" thin layer of nutrient rich water is pumped through the system to provide nutrients and water to the plant.

Processes

Plants uptake essential elements from the soil through their roots and from the air (mainly consisting of nitrogen and oxygen) through their leaves. Nutrient uptake in the soil is achieved by cation exchange, wherein root hairs pump hydrogen ions (H+) into the soil through proton pumps. These hydrogen ions displace cations attached to negatively charged soil particles so that the cations are available for uptake by the root. In the leaves, stomata open to take in carbon dioxide and expel oxygen. The carbon dioxide molecules are used as the carbon source in photosynthesis.

Functions of Nutrients

Although nitrogen is plentiful in the Earth's atmosphere, relatively few plants engage in nitrogen fixation (conversion of atmospheric nitrogen to a biologically useful form). Most plants therefore require nitrogen compounds to be present in the soil in which they grow.

Plant nutrition is a difficult subject to understand completely, partially because of the variation between different plants and even between different species or individuals of a given clone. Elements present at low levels may cause deficiency symptoms, and toxicity is possible at levels that are too high. Furthermore, deficiency of one element may present as symptoms of toxicity from another element, and vice versa.

Carbon and oxygen are absorbed from the air, while other nutrients are absorbed from the soil. Green plants obtain their carbohydrate supply from the carbon dioxide in the air by the process of photosynthesis. Each of these nutrients is used in a different place for a different essential function.

Macronutrients (Derived from Air and Water)

Carbon: Carbon forms the backbone of many plants biomolecules, including starches and cellulose. Carbon is fixed through photosynthesis from the carbon dioxide in the air and is a part of the carbohydrates that store energy in the plant.

Hydrogen

Hydrogen also is necessary for building sugars and building the plant. It is obtained almost entirely from water. Hydrogen ions are imperative for a proton gradient to help drive the electron transport chain in photosynthesis and for respiration.

Oxygen

Oxygen by itself or in the molecules of H_2O or CO_2 are necessary for plant cellular respiration. Cellular respiration is the process of generating energy-rich adenosine triphosphate (ATP) via the consumption of sugars made in photosynthesis. Plants produce oxygen gas during photosynthesis to produce glucose but then require oxygen to undergo aerobic cellular respiration and break down this glucose and produce ATP.

Macronutrients (Primary)

Phosphorus

Phosphorus is important in plant bioenergetics. As a component of ATP, phosphorus is needed for the conversion of light energy to chemical energy (ATP) during photosynthesis. Phosphorus can also be used to modify the activity of various enzymes by phosphorylation, and can be used for cell signalling. Since ATP can be used for the biosynthesis of many plant biomolecules, phosphorus is important for plant growth and flower/seed formation. Phosphate esters make up DNA, RNA, and phospholipids. Most common in the form of polyprotic phosphoric acid (H_3PO_4) in soil, but it is taken up most readily in the form of H_2PO_4. Phosphorus is limited in most soils because it is released very slowly from insoluble phosphates. Under most environmental conditions it is the limiting element because of its small concentration in soil and high demand by plants and microorganisms. Plants can increase phosphorus uptake by a mutualism with mycorrhiza. A Phosphorus deficiency in plants is characterized by an intense green colouration in leaves. If the plant is experiencing high phosphorus deficiencies the leaves may become denatured and show signs of necrosis. Occasionally the leaves may appear purple from an accumulation of anthocyanin. Because phosphorus is a mobile nutrient, older leaves will show the first signs of deficiency.

It is useful to apply a high phosphorus content fertilizer, such as bone meal, to perennials to help with successful root formation.

Potassium

Potassium regulates the opening and closing of the stomata by a potassium ion pump. Since stomata are important in water regulation, potassium reduces water loss from the leaves and increases drought tolerance. Potassium deficiency may cause necrosis or interveinal chlorosis. K^+ is highly mobile and can aid in balancing the anion charges within the plant. It also has high solubility in water and leaches out of rocky or sandy soils. This water solubility can result in potassium deficiency. Potassium serves as an activator of enzymes used in photosynthesis and respiration Potassium is used to build cellulose and aids in photosynthesis by the formation of a chlorophyll precursor. Potassium deficiency may result in higher risk of pathogens, wilting, chlorosis, brown spotting, and higher chances of damage from frost and heat.

Nitrogen

Nitrogen is an essential component of all proteins. Nitrogen deficiency most often results in stunted growth, slow growth, and chlorosis. Nitrogen deficient plants will also exhibit a purple appearance on the stems, petioles and underside of leaves from an accumulation of anthocyanin pigments. Most of the nitrogen taken up by plants is from the soil in the forms of NO_3^-, although in acid environments such as boreal forests where nitrification is less likely to occur, ammonium NH_4^+ is more likely to be the dominating source of nitrogen. Amino acids and proteins can only be built from NH_4^+ so NO_3^- must be reduced. Under many agricultural settings, nitrogen is the limiting nutrient of high growth. Some plants require more nitrogen than others, such as corn (*Zea mays*). Because nitrogen is mobile, the older leaves exhibit chlorosis and necrosis earlier than the younger leaves. Soluble forms of nitrogen are transported as amines and amides.

Macronutrients (Secondary and Tertiary)

Sulphur

Sulphur is a structural component of some amino acids and vitamins, and is essential in the manufacturing of chloroplasts. Sulphur is also found in the Iron Sulphur complexes of the electron transport chains in photosynthesis. It is immobile and deficiency therefore affects younger tissues first. Symptoms of deficiency include yellowing of leaves and stunted growth.

Calcium

Calcium regulates transport of other nutrients into the plant and is also involved in the activation of certain plant enzymes. Calcium deficiency results in stunting. This nutrient is involved in photosynthesis and plant structure. Blossom end rot is also a result of inadequate calcium.

Magnesium

Magnesium is an important part of chlorophyll, a critical plant pigment important in photosynthesis. It is important in the production of ATP through its role as an enzyme cofactor. Magnesium deficiency can result in interveinal chlorosis.

Silicon

In plants, silicon strengthens cell walls, improving plant strength, health, and productivity. Other benefits of silicon to plants include improved drought and frost resistance, decreased lodging potential and boosting the plant's natural pest and disease fighting systems. Silicon has also been shown to improve plant vigor and physiology by improving root mass and density, and increasing above ground plant biomass and crop yields. Although not considered an essential element for plant growth and development (except for specific plant species - sugarcane and members of the horsetail family), silicon is considered a beneficial element in many countries throughout the world due to its many benefits to numerous plant species when under abiotic or biotic stresses. Silicon is currently under consideration by the Association of American Plant Food Control Officials (AAPFCO) for elevation to the status of a "plant beneficial substance".

Silicon is the second most abundant element in earth's crust. Higher plants differ characteristically in their capacity to take up silicon. Depending on their SiO_2 content they can be divided into three major groups:

- Wetland graminae-wetland rice, horsetail (10–15%)
- Dryland graminae-sugar cane, most of the cereal species and few dicotyledons species (1–3%)
- Most of dicotyledons especially legumes (<0.5%)
- The long distance transport of Si in plants is confined to the xylem. Its distribution within the shoot organ is therefore determined by transpiration rate in the organs
- The epidermal cell walls are impregnated with a film layer of silicon and effective barrier against water loss, cuticular transpiration rate in the organs.

Si can stimulate growth and yield by several indirect actions. These include decreasing mutual shading by improving leaf erectness, decreasing susceptibility to lodging, preventing Mn and Fe toxicity.

Micro-nutrients

Some elements are directly involved in plant metabolism (Arnon and Stout, 1939). However, this principle does not account for the so-called beneficial elements, whose presence, while not required, has clear positive effects on plant growth. Mineral elements that either stimulate growth but are not essential, or that are essential only for certain plant species, or under given conditions, are usually defined as beneficial elements.

Iron

Iron is necessary for photosynthesis and is present as an enzyme cofactor in plants. Iron deficiency can result in interveinal chlorosis and necrosis. Iron is not the structural part of chlorophyll but very much essential for its synthesis. Copper deficiency can be responsible for promoting an iron deficiency.

Molybdenum

Molybdenum is a cofactor to enzymes important in building amino acids. Involved in Nitrogen metabolism. Mo is part of Nitrate reductase enzyme.

Boron

Boron is important for binding of pectins in the RGII region of the primary cell wall, secondary roles may be in sugar transport, cell division, and synthesizing certain enzymes. Boron deficiency causes necrosis in young leaves and stunting.

Copper

Copper is important for photosynthesis. Symptoms for copper deficiency include chlorosis. Involved in many enzyme processes. Necessary for proper photosythesis. Involved in the manufacture of lignin. Involved in grain production. It is also hard to find in some conditions.

Manganese

Manganese is necessary for photosynthesis, including the building of chloroplasts. Manganese deficiency may result in colouration abnormalities, such as discoloured spots on the foliage.

Sodium

Sodium is involved in the regeneration of phosphoenolpyruvate in CAM and C4 plants. It can also substitute for potassium in some circumstances.

Essentiality

- Essential for C4 plants rather C3
- Substitution of K by Na: Plants can be classified into four groups:
  1. Group A—a high proportion of K can be replaced by Na and stimulate the growth, which cannot be achieved by the application of K

2. Group B—specific growth responses to Na are observed but they are much less distinct
3. Group C—Only minor substitution is possible and Na has no effect
4. Group D—No substitution is occurred

- Stimulate the growth—increase leaf area, stomata, improve the water balance
- Na functions in metabolism
  1. C4 metabolism
  2. Impair the conversion of pyruvate to phosphoenol-pyruva
  3. Reduce the photosystem II activity and ultrastructural changes in mesophyll chloroplast
- Replacing K functions
  1. Internal osmoticum
  2. Stomatal function
  3. Photosynthesis
  4. Counteraction in long distance transport
  5. Enzyme activation
- Improves the crop quality e.g. improve the taste of carrots by increasing sucrose

Zinc

Zinc is required in a large number of enzymes and plays an essential role in DNA transcription. A typical symptom of zinc deficiency is the stunted growth of leaves, commonly known as "little leaf" and is caused by the oxidative degradation of the growth hormone auxin.

Nickel

In higher plants, Nickel is absorbed by plants in the form of Ni^{2+} ion. Nickel is essential for activation of urease, an enzyme involved with nitrogen metabolism that is required to process urea. Without Nickel, toxic levels of urea accumulate, leading to the formation of necrotic lesions. In lower plants, Nickel activates several enzymes involved in a variety of processes, and can substitute for Zinc and Iron as a cofactor in some enzymes.

Chlorine

Chlorine, as compounded chloride, is necessary for osmosis and ionic balance; it also plays a role in photosynthesis.

Cobalt

Cobalt has proven to be beneficial to at least some plants, but is essential in others, such as legumes where it is required for nitrogen fixation for the symbiotic relationship it has with nitrogen-fixing bacteria. Vanadium may be required by some plants, but at very low concentrations. It may also be substituting for molybdenum. Selenium and sodium may also be beneficial. Sodium can replace potassium's regulation of stomatal opening and closing.

1. The requirement of Co for N_2 fixation in legumes and non-legumes have been documented clearly
2. Protein synthesis of Rhizobium is impaired due to Co deficiency
3. It is still not clear whether Co has direct effect on higher plant.

Aluminium

- Tea has a high tolerance for Al toxicity and the growth is stimulated by Al application. The possible reason is the prevention of Cu, Mn or P toxicity effects.
- There have been reports that Al may serve as fungicide against certain types of root rot.

Crop Disease and Management

History and Importance of Plant Bacteriology

History: History of bacteriology started when Anton Von Leeuwenhoek, (Dutch Worker) observed Bacteria for the first time in 1675 with Microscope. In 1683 he described the bacteria seen with microscope.

The importance of these observations were realised only after Pasteur (1876) had demonstrated their role in fermentation and decay.

Robert Koon (1876) from Germany had proved that the bacteria can cause such disease as anthrax, tuber culosis or Asiatic cholera. He gave four famous Koch's Postulates for proving that a particular organism is the cause of particular disease.

Woronin (1876) isolated and described the root nodule bacteria in leguminous plant. Now it was well recognised that all the organisms' bacteria are most closely related to human being life.

T.J Burbil from University of Illinois (1878) was first proved to association of bacterium with plant disease. He showed that Erwinia amylovora causes fire blight pear and apple.

In 1879 Prilleaux reported the bacterial decay of wheat kernels.

In 1883 J.H Wakker published the results on his through investigationson yellow slime disease of hyacinth caused by bacterium.

J.C Arther (From USA) during 18851887 confirmed Burrill's work.

Luigi Savastano (From Italy) described in 1887 a bacterial knot disease of olives.

Erwin F. Smith (From USA) started studying bacterial disease of pants from 1890. His humerous and excellent contributions on to the study of bacterial disease of plants, particularly of the bacterial wilts of cucurbits, of solanaceous crops, and of cruifers established beyond any doubt the role of bacteria as phytopathogens. He resolved the controversy with the German Bacteriologist, Alfred Fischer (1897,1899) who did not think that bacteria were the primary cause of disease in plants. He also was among the first to notice (1893-1894) and to study the crown gall disease. He showed it to be caused by bacteria, studied its anatomy and development, and considered it to resemble cancerous tumors of human and animals.

Chilton et al. (1977) showed that the crown gall bacterium transforms normal plant cells into tumor cells by introducing into them a plasmid, part of which becomes inserted into the plant cell chromosomes DNA.

Windsor and Black in 1972 observed rickettsia like organisms in the phloem of clover plants infected with the club leaf disease. The following year similar organisms were observed in grape infected with pierces disease, in peach infected with phony peach, and others.

E.J Butler from India, examined the brown rot disease of potato in 1903, and indicated the bacterial nature of the disease.

M.K Patel and associates from Pune, India, commensed a serious of studies on bacterial phytopathogens in 1948. This team, over a period of about 15 years, reported on nearly 40 bacterial diseases, some of them on economically important host plants.

Importance: About 1600 bacterial species are known. Most are strictly saprophytes which are beneficial to human because they decompose the enormous quantities of organic matter produced yearly by humans, animals and factory waste products or by death of plants and animals and factory waste products or by death of plant and animals. Several species causes disease in humans, animals and plants.

General Characters of Plant Pathogenic Bacteria

1. Almost all plant pathogenic bacteria are rod shaped, the only exception being Streptomyces, which is filamentous.
2. The rod shaped bacteria are more or less short and cylindrical and in young cultures, they range from 0.6 to 3.5 m in length and from 0.5 to 1.0 m in diameter.
3. In older cultures or at high temperatures, the rods of some species are much longer and they may even appear filamentous.
4. Sometimes deviations from the rod shape in the form of a club , a Y or V shaped, and other branched forms occur, and some bacteria may occasionally occur in pairs or in short chains.
5. The cell walls of bacteria of most species are enveloped by a viscous, gummy material, which may be thin (Slime layer) or may be thick, forming a relatively large mass around the cell (Capsule).
6. Most plant pathogenic bacteria are equipped with delicate, thread like flagella.
7. In some bacterial species each bacterium has only the flagellum, others , have a tuff of flagella at one end of the cell (Polar flagella); some have a single flagellum or a tuft of flagella at each end, and till others have peritrichous flagella, that is, distributed over the entire surface of the cell.
8. In the filamentous Streptomyces species, the cells consist of non septate branched threads, which usually have a spiral formation and produce conidia in chains on aerial hyphae.
9. Single bacterium appears hyaline or yellowish white under the compound microscope.
10. Bacteria grow and produce colonies on solid medium.
11. Colonies of different species may vary in size, shape, form of edges, elevation and colour, and are sometimes characteristics of a given species.
12. Bacterial cells have thin, relatively tough, and somewhat rigid cell walls.
13. All the material inside the cell wall constitutes the protoplast.
14. The nuclear material consist of a large circular chromosome composed DNA and appear as spherical, ellipsoidal or dumbbell shaped body within the cytoplasm.
15. Often bacteria also have single or multiple copies of addition smaller circular chromosomes called 'Plasmids' that can move

or be moved between bacteria or between bacteria and plants as for example in the crown gall disease.

16. Rod shaped Phytopathogenic bacteria reproduce by the asexual process known as binary fission or fission. Under favourable conditions bacteria may divide every 20 minutes.
17. Almost all plant pathogenic bacteria develop mostly in the host plant as parasites and partly in plant debris or in the soil as saprophytes.

Classification of Phytopathogenic Bacteria

According to the 8th Ediction of Bergy's Manual

Phytopathogenic bacteria are classified as:

Division II: Sctobacteria (indifferent to light)

Class I: The Bacteria

Part: Gram negative aerobic rods and cocci.

Family: Pseudomonaceae

Genera: 1. Pseudomonas

Pseudomonas:

Characters: Cells single, straight or curved rods but not helicie, generally 0.5 X 1.5 to 4 microns in size, motile by polar flagella one or more in number, non spore forming, gram negative; strict aerobes ; chemoorganotrophs; metabolism respiratory, never fermentative.

Species:

1. Pseudomonas solanacearum causes brown rot or wilt of potatoes.
2. Pseudomonas rubrillineans causes red stripe of sugarcane.
3. Xanthomonas:

Characters: Cells single, straight rods measuring 0.2 to 0.8 X 0.6 to 2.0 microns (Usually 0.4X 1.0 microns); Gram –ve , motile by a single polar flagellum; non spore forming; copious extracellular slime produced; strict aerobes; metabolism respiratory, never fermentative ; oxidase negative and catalase positive; bacterial colonies yellow coloured due to production of Xanthomonadin pigment.

Species:

1. X. campestris pv. Citri causes citri cause citrus canker.
2. X. campestris pv. Malvacearum causes angular leaf sport and black arm of cotton.

3. X. campestris pv. Oryzae causes bacterial leaf blight of rice.

Family: Rhizobiaceae

Genus: Agrobacterium

Character: The bacteria are rod shaped, 0.8 X 15. m. They are motile by means of 1.4 peritrichous flagella, when only one flagellum is present it is more often lateral than polar. When growing on carbohydrate – containing media the bacteria produce abundant polysaccharide slime. The colonies are non-pigmented and usually smooth. These bacteria are rhizosphere and soil inhabitants.

Species: Agrobacterium tumefaction causes crown gall of pome and stone fruit trees, brambles and grapes.

A. rubi causes cane gall of raspberries and black berries.

A. rhizogenes causes hairy root of apple.

Gram Negative, Facultative Anaerobic Rods

Family: Enterobacteriaceae

Genus: Erwinia

Characters: Cells predominantly single, straight rods, measuring 0.5 to 1.0 X 1.0 to 3.0 microns; motile (Except E.stewartii and E.dissolves) by peritrichous flagella; gram negative ; grows well on artificial media aerobically as well as anaerobically and produce acid. The genus has three major groups of species.

i) Pectolytic bacteria that cause oft rot.

ii) Those do not cause soft rot but dry necrosis and wilt and

iii) Epiphytes that cause neither soft rot nor necrosis.

Species:

1. E. amylovora causes fire blight of apple.
2. E. tracheiphia causes vascular wilt of cucurbits.
3. E. carotovora pv. atroseptica causes black leg of potato.
4. E. carotovora pv. Carotovora causes soft rot numerous fleshy fruits, vegetables and ornamentals.

Gram Positive, Irregular Rods and Filamentous Bacteria

Irregular Rods: Corynebacterium (Plant Pathogenic)

Genera:

1. Curtobacterium
2. Clavibacter

Characters of Curtobacterium: Cells small short rods, coccoid cells found in old cultures, weakly gram negative, frequently old cell lose Gram Positivity generally motile by lateral flagella; cell multiplication by bending type of cell division; pleomorphism only slight; major cell wall amino acid is ornithine; G + C content of the DNA ranges form 66 to 71 moles per unit.

Species: Curtobacterium flaccumfaciens pv. Flaccumfaciens (Hedges) Collins and Jones (Corynebacterium flaccumfaciens Dowson) causes vascular wilt of beans.

Characters of Clavibacter: The cells are Gram positive; non avoid fast plemorphic rods, often arranged at an angle to give V-formation as a result of snapping or bending type of cells division, no coccoid cells are seen; non endospore forming; non motile; strict aerobes; nutritionally exacting; nitrate not reduced; cell wall peptidoglycan contains diaminobutyric acid; G+C content of the DNA is 70 +-5 moles per cent.

Species: Clavibactor michinganese subspecies michiganese (Smith) Davis et al. (Corynebacterium michiganese or C. michiganese pv. michiganese or C. michiganese sub species michiganese causes canker of tomato and chilli.

Filamentous Bacteria:

Order: Actinomycetales

Family: Streptomycetaceae

Genus: Streptomyces

Characters: Slender, coenocytic filaments, 0.5 to 2.0 microns in diameter; aerial mycelium at maturity forms chains of three to many spores; cell walls contain diaminopimelic acid; Gram Positive; aerobic; heterotrophs; generally reduce nitrates, on isolation colonies are small, 1 to 10 mm in diameter, descrete and lichenoid , leathery or butyrous, initially relatively smooth but later develop a weft of aerial mycelium that may appear granular, powdery, velvety or floccosed.

Species: S. scabis causes common scab of potato.

Symptom Caused by Phytopathogenic Bacteria

Local Lesions: Localised spots may occur on the leaf blade, petiole, stem and fruits. The leaf spots starts as minute water soaked specks, spreading rapidly to form circular, irregular or angular spots, bound by veins and veinlets. The pathogen which enters the host tissues through stomatal openings establishes in the substomatal space and remains so in the parenchymatous tissues causing necrotic lesions.

E. g Angular leaf sop of cotton caused by Xanthomonas campestris pv. malvacearum.

Blights: The invasion by the bacterium loads to very rapid and extensive necrosis of the affected plant parts resulting in scorched appearance of the host.

E. g Bacterial blight of paddy caused by Xanthomonas campestris pv. oryzae.

Soft Rots: The soft rot symptoms are found mainly on fleshy parts. The major effect is softening of the tissues due to disintegration of cells and dissolution of middle lamella, as a result of enzyme action. Very often a dirty liquid oozes out of the affected parts. In many causes, the disintegration is preceded by change of colour.

E.g Black leg and soft rot of potato caused by Erwinia carotovora sub species carotovora and Erwinia carotovora sub species atroseption.

Tumors and Galls: In many bacterial diseases the effect of invasion by the pathogen is hyperplasia and hypertrophy of invaded tissues. As a result, tumours develop on the affected organs.

E. g Crown gall tumors on rose plant caused by Agrobacterium tumefaciens.

Vascular Disease: In same of the bacteria leaf spot diseases the organism moves into the vascular system and becomes systemic. In others, the invasion is concentrated in the vascular tissues causing typical wilt of the plant by plugging the water conducting vessels and by production of toxins.

E.g Bacterial canker and wilt of tomato caused by Clavibactor (Corynebacterium) michiganese sub species michiganese. Bacterial wilt of tomato, potato, tobacco and egg plant caused by Pseudomonas solanacearum.

Scabs and Cankers: Scabs and cankers are corky outgrowth which is formed on leaves, twigs and all other plant parts above the ground. These outgrowths are the result of the reactions of the host tissues to the pathogen. Such reactions are mostly localised and often confirmed to the parenchymatous tissues of the host plant. Scab is formed by epidermal infection and is not deep seated. Cankers on the other hand, are deep seated and involve the cambium layer.

E. g 1. Citrus canker caused by Xanthomonas Campestris pv. citri.

2. Common scab of potato caused by Streptomyces scabs.

Mode of Entry of Bacteria in Plants

Unlike fungal pathogens, bacteria are incapable of mechanically penetrating the cultinized plant tissues, culticle, periderm, etc. Since the ability to form appresoria is lacking in them. The ways of gaining entry into the plants are as follows.

1. Active invasion by bacteria through non-cultinized surface is sometimes feasible by means of a large mass of cells functioning together. Rhizobium spp. Which causes nodules in legume form a muciliagenous bacterial colony or zoogloea at the apex of the root hairs or root tips, the Zoogloea dissolves the cell wall and then, in the form of a curved and branched infection filament or infection thread advances into the interior of root hair and hence, into the root corner.
2. The cells of black rot of caggage, Xanthomonas campestris, enter through hydathodes, which are specialized gland cells at leaf edge.
3. The fire blight bacterium, Erwinia amylovora, enters the floral parts through the special nectar producing cells present system.
4. The potato scab organism, Streptomyces scabies, gains entry into the tubers through lenticels.
5. Entry through stomata: The mechanism of entry into the stomato is through water congestion which leads the cells into the open stomato assisted by wind blown rain or dew.

e.g 1. Pseudomonas tabaci causing wild fire of tobacco.

2. Xanthomonas. campestris pv. malvacearum causing angular leaf spot of cotton.
3. Xanthomonas. Phaseoli causing bean blight.
4. Xanthomonas. Campestris causing black rot of cabbage.

6. Entry through Wounds:
   - During various cultivation operations and while harvest, transport and storage, injuries to various plants parts are caused. Many bacteria present in vicinity enter the host tissue.
   - E.g Erwinia spp. causing soft rot of vegetables.
   - Agrobacterium tumefaciens causing crown gall of fruit trees.
7. Entry through Insect Wounds:

E. g Crown gall of fruit trees and wilt of cucumber.

8. Entry through wounds caused by plant parasitic nematodes and funi. Root disease complex in citrus.

Important Phytopathogenic Bacteria

Sr.No	*Genus*	*Family*	*Gram Reaction*	*Characters*	*Symptoms produced on host*	*Example*
1	Pseudomonas	Pseudomonadeveae	-Ve	1. Rod straight to curved, size 0.5-1 X 1.5-4 mue m, one to many polar flagella.2.Colonies not yellow, do not produce acid from lactose.	Leaf spot, blights, vascular wilts, soft rots, canker.	1. Brown rot or bacterial wilt of potato.2 Bacterial wilt of brinjal and tomato.
2	Xanthomonas	Pseudomonadeveae	-Ve	1. Straight rods, size 0.4-1X 1.2 -3m. One polar flagellum present.2. Colonies yellow due to Xanthomonadin , produce acid from lactose.	Leafspot, fruit spots, blight, canker	1. Citrus canker, 2. Black arm of cotton,3.Blight of paddy.
3	Agrobacterium	Rhizobiaceae	-Ve	1. Rod shaped, size 0.8 X 1.5 -3 mue m. 1 to 4 peritrichous flagella.2. Colonies mostly white do not hydrolyze starch.	Crown Gall	Crown Gall of stone fruit roses.
4	Erwinia	Enterobacteriaceae	-Ve	1. Straight rods, size 0.5-1, 0-3.0 mue m , several peritrichous flagella.	Fire blight, wilt soft rot.	1. Fire blight of apple.2. Soft rot of vegetables.
5	Clavibactor (Corynebacterium)	Enterobacteriaceae	+Ve	Straight to straightly curved rods, 0.5-0.9X 1.5 - 4.0Mue m. non motile but some species are motile by one or two polar flagella.	Wilt	Wilt of potato ad Tomato
6	Streptomyces	Streptomycetaceae	+Ve	Slender branched hyphae without	Scab	Potato scab

Perpetuation of Bacteria

The Phytopathogenic bacteria do not produce any resting structures. They survive by some of the following ways:

Survival on Self Sown Plant, Collateral Hosts: Bacterial pathogen survives and multiplies on self sown plants of host crop, collateral hosts during the main and the off seasons. The collateral hosts are most often perennial or noxious weeds growing in the host crop fields, or in their vicinity while others may be cultivated economic crop plants. The host range of some bacterial plant pathogens is very wide.

E. g Pseudomonas solaneceanum -200 host plants.

Agrobacterium tumefaciens- 150 host plants.

Survival on Host Crop: The pathogens of the bacterial leaf blight and the leaf streak of rice may survive from crop to crop in double or triple – rice culture areas in the tropics. The latent infections as in cankers, galls, bud scales, the cracks in the bark or sheltered place are often further protected by the host mucilage in trees as in the citrus canker or the fire blight.

Survival on Non Host Plants: Xanthomonas campestris pv. Citri causal agent of citrus canker has been found to survive in the non-host plants. The epiphytic (Saprophytic) existence of X.campestris pv . phasebli on the phyllosphere of Phaseolus vulgaris has been found.

Survival Through Crop Residues and in Soil: Some of the Phytopathogenic bacteria are capable of living saprophytically on plant residues or parennate in the soil in Free State or on any organic matter. E.g Erwinia spp. Pseudomonas, solanacearum racee2, Xanthomoans campestris pv. malvacearum. Most of the bacterial pathogens of aerial parts can not survive in a natural soil for a long time due to their poor competitive saprophytic ability and antagonism of other microorganisms.

Survival Through Seeds and Vegetative Propagation Parts: Some of the phyopathogenic bacteria survive in seeds of some crops and vegetative plant parts used for propagation E. g Xanthomonas campestris pv. oyzae and Xanthomonas compestris pv. malvacearum survive through seeds. Pseudomonas solanacearum survive in potato tubers.

Transmission (Dissemination) of Bacteria

The bacterial diseases are mainly disseminated through the agency of seed, air and water.

Seed: Xanthomonas campestris pv. oryzae (in Paddy), X.campestris pv. malvacearum (in cotton) are carried either externally or internally through seed. Corynebacterium sepedonicum and pseudomonas solanacearum (in sugarcane) are spread through vegetative plant parts used for propagation.

Soil: Phytopathogenic bacteria live as saprophytes on plant residues in the soil and when suitable host plants are sown they become active pathogens. They are disseminated from place to place by means of cultivation operations, tool of cultivation and related agencies- human and animal and also through irrigation water, rain water and wind E.g of disease spread: Bacterial blight of cotton (X. campestris pv. malvacearum).

Wind: Bacteria surviving on fallen leaves etc and in soil are carried in the wind for long distance and deposited on the host plants.

Rain and Water: Rain drops falling on bacterial pustules or oozo present on the host surface distribute the bacteria in fine droplets to to other portions of the plant. The motile ones can move short distances in drops of water on plant parts and gain entry into the plant tissues. Rain water especially when accompanied by wind, forms an important agent for rapid spread of the pathogens. Bacteria in soil are carried via rain or irrigation water. E.g of disease spread: leaf blight of rice Xanthomonas campestris pv. oryzae.

Insects: Insects play a major role in the dissemination of bacteria. They act as carries of bacterial cells as well as help in infection through wounds E.g disease spread through insects: Fire blight of apple (Erwinia amylovora).

Mechanical: Use of contaminated tools and bags during cultural operations and movement of infective material by implements and other agencies during various field operations also play an important role in the spread of bacterial disease e.g of disease spread : Brown rot of potato (Pseudomonas solanacearum)

Man: Man is the agent behind the long distance dissemination which takes place chiefly through infected seeds and vegetative propagating material. E. g of disease spread: Leaf blight of paddy (Xanthomonas. compestris pv. oryzae), brown rot of potato (Psudomonas solanacearum).

Management of Bacterial Diseases

Bacterial plant pathogens directly enter the host tissue either through natural openings or through wounds; therefore, it is difficult to manage them. Some of the general principles of managing the bacterial disease are as follows:

Exclusion: Exclusion of a disease entry either from a foreign nation or within the country, fro a diseased to a healthy tract is an effective measure for avoiding the disease. Quarantine measures are in vogue in several countries. E .g diseased managed: Citrus canker (Xanthomonas campestris pv. citri) in U.S.A.

Eradication: Eradication is generally carried out for eliminating a well established pathogen and its host plant, collateral hosts, or insect vectors and some times use of cultural measures for starving out or killing the pathogen E. g of disease managed.

i) The complete distruction of all infected trees was successfully achieved the citurs canker in Florida, USA during 1914-1927.

ii) Pruning of the infected twigs is followed to reduce the inoculums in orchards for citrus canker.

iii) The destruction of volunteer plants and weed hosts brings down the inoculum level. e.g Brown rot of potato (Pseudomonas Solanacearum).

iv) The crop residue may be burnt or ploughed deep into the soil with watering to ensure decomposition, which is helpful for the pathogens which cannot live saprophytically in the soil. E. g Bacterial blight of cotton (Xanthomonas campestris pv. malvacearum).

v) Crop rotation with cereals is advocated for the management of wilt of tobacco (Pseudomonas solanacearum).

vi) The selected application of NPK , particularly low and split doses of N, helps to reduce the intensity of the bacterial leaf blight of rice (Xanthomonas campestrsi pv. Oryzae).

vii) The control of insects is helpful in reducing the soft rot of vegetables (Erwinia spp.) and the citrus canker (Xanthomonas campestris pv. citri).

viiii) The avoidance of cultural mismanagement favourable to disease is important such as water logging in the nurseries (against the bacterial leaf blight of rice), flooding or over irrigation in the field. (against the soft rot disease)

ix) The sterilization of the cutting knife by flame or by 0.1 % KMNO4 solution while cutting the potato tuber for sowing is recommended against the potato wilt.

x) The seed certification programmes to raise pathogen free seeds. E.g bean (Xanthmonas campestris pv. campestris) are in practise in U.S.A.

xi) Elimaknation of the extremally and internally seed born pathogen by seed treatment.

E. g 1. Delinting of cotton seed with concentrated sulphuric acid for bacterial blight of cotton.

2. Hot water treatment of cotton seed at 560C for 10 minutes for bacterial blight of cotton.

3. Soaking of rice seeds in 0.025% strpetocyclic solution for leaf streak disease of rice.

xii) When tobacco is immediately grown after maize there is a considerable reduction in the incidence of Pseudomonas solanacearum on the later host.

xiii) There is some prospect of biological control with the application of organic matter in the form of compost and green manure and even inoculation of antibiotic micropropaganism, encouraging antagonism by the application of superphosphate, application of phages in seed plant and soil as well as by inoculating with bacteriocin- producing strains.

Plant Sanitation: These measures are adopted so as to avoid the disease onset and also to prevent spread.

E.g. i) Removal of dead plant parts from fruit crops and protection of cut surface with suitable bactericides (E. g for citrus canker and fire blight of apple).

ii) Avoiding injuries to the plant parts at the time of cultural operations and during harvest, transport and storage.

iii) Collection and destruction of diseased fallen leaves, bolls etc. for bacterial blight of cotton.

Plant Protection: These measures are adopted to prevent the onset and subsequent spread of plant diseases.

E. g i) Seed treatment of cotton and rice with antibacterial chemicals against bacterial blight and leaf streak diseases respectively.

ii) Treatment of vegetatively propagation material with bactericides e.g 0.02% stroptocycline for 30 minutes treatment of seed tubers of potato against brown rot disease.

iii) Foliar sprays with Bordeaux mixture and copperoxy – chloride against leaf spots and blights.

iv) Foliar sprays of streptomycin sulphate, 100 and 500 ppm against fire blight of apple and citrus canker.

v) Therapy: Pancillin and vancomycin are effective in the disintegration of the crow gall. The application of antibiotics for protection also involves therapy. The bacteriocines have also been demonstrated to be successful against bacterial disease E. g Agrobacterium radiobacter against crown gall of peach and tomato seedlings (Agrobacterium fumetaciens.)

vi) Immunisation: By screening under artificial epiphytic conditions, resistance source for bacterial pathogens can be known. Resistant varieties are evolved by selection, breeding and other methods. E. g HC-9, BJA-592, P-14, T-12, 101-102 B, Reba-B-50, Khandwa-2, DHY-286, B-1007, cultivars of cotton and N-22,IR-22 cultivars of rice resistant to bacterial blight disease of these crops.

vii) Disease Forecasting: Plant disease forecasting help in timely and economic management of bacterial diseases. Following basic principles are used for forecasting.

a. Forecasting by natural infection: This is based on observation of plants at regular intervals- both in the nursery as well as in the main field for the occurrence of the disease.

b. Forecasting by climatic condition: Effect of various metrological parameters on the incidence and spread of the disease in studied in detail and the congeneal meteorological parameters for these are workedout. Prediction model is prepared on the basis of these observations.

c. Forecasting by bacteriophage population: A correlation has been obtained between the occurrence of bacteriophage in the field (Soil, irrigation water, etc.) and disease outbreak in the case of bacterial leaf blight of rice. Prior to outbreak of the disease the phage population in soil and irrigation water increase. It is possible, to some extent to forecast the disease outbreak using the phage population as an index.

Definition and Characteristics of Virus

Definition of Virus: Mathwas (1981) considers a virus as a set of one or more template molecules normally encased in a protective coat or coats of protein or lipoprotein, which is able to organize its own replication only within suitable host cells where its production is:

i) Dependent on hosts protein synthesizing machinery (ribosomes).

ii) Organised from pools of required material rather than binary fission and

iii) Located at sites which are not separated from the host cell contents by a lipoprotein bilayer membrane.

Bos (1983) defines virus as an infectious agent often causing disease, invisible with the light microscope (Sub-microscopic) , small enough to pass through a bacterial filter, lacking a metabolism of its own and depending on a living host cell for multiplication. Viruses are small packages of host alien genetic information of one type. (RNA or DNA), either in one strand or in a few segment. Encapsulated together or separately and enclosed in a coat of one or more types of protein, some time with an extra coat (envelope) and some other constituents.

Characteristic of Virus: Following characters will always be includes:

i) Viruses contain one or more pieces of a single type of nucleic acid, either RNA or DNA, never both.

ii) The nucleic acid carries the genome of the virus which differs from one virus to another.

iii) The genome in the nucleic acid strand directs the synthesis of specific proteins for the protein coat which must be present in all viruses throughout their active phase except at the time of replication when, protein coat and nucleic acid are separated.

iv) Viruses rely on living host cells for most of the enzymes necessary for their replication.

v) Viruses are in cable of growing for the synthesis of Lipman system is absent in Viruses.

History of Plant Virology

1. Virus disease of plants was known long before the discovery of bacteria.

   i) Breaking of flower colour of tulips (as early as 1576).

   ii) Transmission of leaf variegations from the scion to the stock of woody plants (as early as 1700).

2. Tobacco mosaic was identified by Swietch in Holland in 1857.

3. Adolph Mayer (1886) was the first to point out that tobacco mosaic is readily transmissible and infectious. Believed that bacteria are the cause of the disease.

4. Iwanowski (1892) confirmed some of the results Mayer. He demonstrated that the power to infect was lost if the sap was previously heated. He reported that infectiousness was retained even when sap was passed through bacteria proof filters.
5. Bejierink (1896) demonstrated that tobacco mosaic infectious agent could diffuse through an anger membrane and concluded that tobacco mosaic was caused by a non-corpuscular ' Contagium vivum fluidum' which he called as Virus.
6. Hashimoto (1894) showed transmissible lity of rice dwarf disease by leaf hopper (Nephotettix apicalis var. cincticeps).
7. Stanley (1935) crystallized tobacco mosaic virus with ammonium sulphate and concluded that the virus was an autocatalytic protein that could multiply within the living cells. He considered that virus was a globulin containing no Phosphorus. For his discovery he was awarded a Nobel Prize.
8. Bawden and Pirrie (1936) described the isolation from TMV infected plants of a liquid crystalline nucleoprotein containing nucleic acid of the pentose type. They showed that the particles were rod shaped.
9. Kausche, Pfankuch and Ruska (1939) saw virus particles for the first time with the electron microscope. They confirmed that TMV was rod shaped.
10. Muller (1942): Williams and Wycoff (1944) developed shadow casting technique with heavy metals which was useful for determining the overall size and shape of the virus particles.
11. Markham and Smith (1949) isolated TYMV and showed that it contained (1) An infectious nucleoprotein (about 25%) and (2) non inferious protein.
12. Morel and Martin (1952) showed that virus free plants could be obtained from totally infected parents using meristem tip culture.
13. Kassanis (1954) showed that virus could be eradicated from infected plants by high temperature treatment.
14. Gierer and Schramm (1956) showed that the protein could be removed the virus and that the nucleic acid carried the genetic information so that inoculating with the nucleic

acid alone cause infection and could reproduce the complete virus.

15. Kassais (1962) was the first to describe the satellite virus (Sv) which has been found only in association with tobacco necrosis virus.
16. Doi et.al. (1967) recognised MLO disease (Yellow witches broom).
17. Ishiie et.al. (1967) showed that the MLO bodies and the symptoms disappeared temporarily when the plants were treated with tetracycline antibodies.
18. Black and Markhan (1963); Miura et.al. (1966) showed that wound tamout and rice dwarf viruses contain double started RNA.
19. Shephard et.al. (1968) showed that cauliflower mosaic virus contains double stranded DNA.
20. In 1971, Diener determined that the potato spindle tuber disease was caused by a small (250- 400 bases long), single stranded circular molecule of infectious RNA which he called a viriod.
21. Windsor and Black (1972) observed RLO in the phloems of clover plants infected with clu leaf disease.
22. Davis and Worley (1973) observed motile; helicle microorganism associated with corn stunt disease and named it spiroplasma.

Economic Importance of Plant Viruses

Viruses cause serious disease of crop plants reducing both quality and quantity of final produce. Also keeping quality of plant produce is also affected. Some of the important viral diseases are:

Tobacco mosaic virus, Tomato mosaic, Tomato spotted wilt, Potato leaf roll, Potato virus X and Y in Potato, Papaya mosaic, Citrus tristiza, chilly leaf curl, Banana bunchy top etc.

Nature of Virus: Each plant virus consists of two components the nucleic acid and the protein coat or capsid. The mature particle of a plant virus is generally called virion and the whole infective particle is called as Nuclcocapsid.

Nucleic Acid of Viruses:

1. Nucleic acid portion of the virus particle is called as Genome.
2. It is actual infective component.

3. Majority of plant viruses contain RNA (Ribonucleic acid).
4. Some viruses (Cauliflower mosaic virus, mung bean yellow mosaic virus, maize streak virus) contain DNA (Deoxyribo nucleic acid).
5. The nucleic acid may be in a single strand or in a double strand.
6. Most plant viruses contain single strand of RNA.

These strands may be free at both ends (liner RNA strand) or the end may be jointed together to form a circular nucleic acid molecule (Circular RNA).

7. In double stranded viruses (as RNA or ds DNA) the two strands are called around each other helically.
8. The nucleic acid may be present as a single continuous strand or it may be present as two or more piece in the same or different particles.

The Virus Protein:

1. The virus protein is termed as capsid.
2. It protects nucleic acid (RNA or DNA) of virus.
3. It is made-up of different amino acid sequences in different viruses. This amino acid are linked by Peptide bonds.
4. The protein sub units of the capsid are known as structural units when these units form groups of 2,4,6 are called Morphological units or capsomers.
5. The sequence of amino acids constitutes the primary structure of the protein. Helical structure produced by amino acids is known as secondary structure of the protein and folding of protein gives tirtiory structure.
6. Protein shell or coat protects viral nucleic acid from environment.

Membrane Bound Viruses: The rhabdo such as potato yellow dwarf virus, lettuce necrotic yellows are provided with a outer envelops or membrane bearing surface projections. Inside the membrane is the nucleocapsid.

Satellite Viruses: These are viruses associated with certain typical viruses but depend on the latter for multiplication and plant infedtion and reduce the ability of the typical viruses act like parasite of the associated typical virus.

Satellite RNAs: These are small, liner RNAs found in virions of certain multicomponent viruses.

Viroids: These are small (250-400 nucleotide), naked, single stranded, circular RNAs capable of causing disease in plants by themselves.

Virusoids: These are viriod like, small, single stranded , circular RNAs that are present inside some RNA viruses, virusides are the part of genetic material of these viruses and therefore , form an obligatory association with these viruses so that neither the virus not the virusolcan multiply and infect a plant in the absence of its partner.

Morphology of Viruses

Plant viruses are usually described as

1. Elongated (Rigid Rod or Flexious thread):
   a. Rigid rod – (Ex) Tobacco mosaic virus (15 X300 nm) Barley stripe mosaic virus (20x 10 nm).
   b. Flexuous thread (Ex) Potato virus X 10-13 wide upto 480 nm length , citrus tristeza virus (10-14 X 2000nm).
2. Rhabdo Viruses: Bacillus like- These are short bacillus like rods approximately 3 to 5 times as long as they are wide.

(Ex): Potato yellow dwarf virus 75X 380 nm.

Wheat striate mosaic virus – 65X 270 nm.

Lettuce necrotic yellow virus – 52 X 300 nm.

3. Spherical (Isometric or Polyhedral) :

All spherical viruses are actually polyhedral ranging in diameter about 17 nm to 60 nm.

Ex: Tobacco Necrosis satellite virus 17 nm in diameter.

Wound tumour virus 60nm in diameter.

Tomato spotted wilt virus 70-80 nm in diameter.

Current Category » Crop Disease and Management

Properties of Plant Virus

Biological Properties: Each virus produces its own protein, its function is to protect RNA from host enzymes (Ribonuclease), heat, ultra violet light and chemical protein has no infectivity.

Infectiousness of Virus:

1. Viruses are infectious and highly contagious, infectivity depends on virus synthesis.
2. After the entry of pathogen in host through natural opening or wounds or pollen grains, virus comes in contact with with inoculation.

3. This is done by Host, since viruses do not produce enzymes. They lack the Lipman Enzymatic system for the conversion of high energy into potential energy required for biological activity.
4. So they have to depend on hosts. This is a major difference between the host parasite relationship in viral diseases and those of other pathogen.
5. The naked RNA induces host cell to form enzyme RNA polymerase. These enzyme in presence of viral RNA and nucleotides produce additional RNA. The new viral RNA induces host cell to produce specific protein molecule required for its coat.

Physiological Properties of Virus:

Dilution End Point (DEP): Dilution of Juices containing the virus may or may not be affected the infectivity. Tobacco mosaic virus is reported to remain infectious even in a dilution of 1:1000000 (6) Cucumber mosaic virus retains, Virulence at 1: 1000 Potatorugose mosaic virus causes poor infection when diluted to 1:10 or 1: 100.

Thermal End Point (TEP): Effect temperature, no of viruses can be inactivated by Thermo therpy. Tomato mosaic virus is destroyed by treating virus containing Juice at 85 to 90 degree centigrade for 10 minutes, lower temp. Seems to have no effect on the viability of plant viruses.

Longevity in Vitro (LIV): Retention of infectivity in storage Storability of Juice), if the Juice from mosaic infected tobacco plant is kept; the virus can remain viable for a few hours to several months. TMV in dried leaves or in Juice dried on filter paper remains infective for many years.

Chemical Properties:

Host Range: Host range is useful in distinguishing the viruses from one another. The viruses are inoculated into indicator an plant which develops typical symptoms, local lesions, ring spot systemic symptoms etc.

Ex: EMV and TMV incite symptoms on tobacco but EMV affects cucumber systematically where as TMV does not.

Mutability and Strains: The presence of genetic material in the form of RNA in plant viruses ensure that new strains of the viruses may develop probably by mutation of RNA. In tobacco mosaic virus alone, there are more than 50 strains.

Serological Reactions: If a virus containing juice is injected into body of a rabbit, the rabbits form antibodies, which will react with viral proteins to give precipitation. The reactions are specific. i. e the antibodies obtained by inoculation of strain A, of a virus will precipitate , the juice containing the same specific type of virus.

Distribution and Movement of Viruses in Plant

1. After introduction into cell, virus moves towards site of synthesis. This movement is passive through protoplasmic stream.
2. The movement of synthesised virus particles between cells occurs through plasmodesmata connecting adjacent cells.
3. The rate of spread is greater in young than old tissues. The movement faster at higher than lower temperature due to fast streaming of protoplasm. Some viruses transported through phloem, few through xylem vessels.

Virus and its Antibody are Brought Together in Several Ways

i) Precipitation test: In this antibodies and antigens are mixed together in solution.

ii) Ring interface test: They meet at the interface between 2 solutions containing each separately.

iii) Gel diffusion test: The diffuse towards each other through an agar gel and meet in a zone in suitable concentration.

iv) Agglutination reaction: Sometimes antigen is observed on a surface of large particles, such as a cell, plastid and these are precipitated by addition of antibodies.

Mycoplasma Like Organisms Or Pleuro Pneumonia Like Organisms (PPLO)

Mycoplasmas are the smallest, walless, free living microorganisms, cause disease similar to virus in plants. Doi ct.al. (1967) described the structure of this pathogen using election microscope. Mycoplasmas like bodies have also been detected in electron micro graphs of plants suffering from corn. Stunt and aster yellow.

Definition

"Mycoplasma are smallest, free living, walless, prokaryotic organism without an organised and bounded nucleus."

Mycoplasma lack rigid cell wall, being surrounded only by single tripla layered unit membrane which allows them to be highly plemorphic. They assume vast array of shapes and size. They require sterols (Lipoproteins) for growth Mycoplasma cannot be grown on artificial media and they reproduce by budding and binary fissions.

Taxonomy

Mycoplasma Like Organisms are Included in Class: Mollicutes which has one order: Mycoplasmatales. The order has three families, each family containing one genus.

1. Mycoplasmataceae: One genus Mycoplasma.
2. Acholeplamataceae: One genus Acholeoplasma.
3. Spiroplasmataceae: One genus Spiroplasma.

These genera are separated on the basis of their sterol requirement for growth.

Characteristics of Mycoplasma Like Organisms (MLOs

1. MLOc lack true cell wall and ability to synthesize the substances required to form a cell wall.
2. MLos are bounded only by a single triple layered unit membrane.
3. They are small, sometime ultra-microscopic cell containing micro cytoplasm, randomly distributed ribosome and strands of nuclear material.
4. They measure from 175 to 250 nm in diameter during reproduction but grow into various shapes and size lateron.
5. Shapes range from coccid or slightly ovoid to filamentous, sometimes, they produce branched myceloid structure.
6. They are capable of reproducing by budding and binary transverse fission on coccid and filamentous cell.
7. MLOs have no flagella, produce no spore and are gram-ve.
8. Nearly all MLOs are parasitic to humans and animals and all saprophytic. Once can be grown on more or less complex, Artificial nutrient media in which they produce minute, colonies , that usually have a characteristic "fried egg" appearance.
9. MLOs have been isolated mostly from healthy and diseased animals and human suffering from the disease of respiratory and urogential tracts (associated with arthritic and nervous disorder of animals).
10. Most MLOs are completry resistance to penicillin, but they are sensitive to tetracyclic, chloramphenicol, some to erythromycin and to certain other anti biotics.

Mycoplasma Like Organism of Plant

The organism observed in plants and insect vector with the exception of Spiro plasma, resemble the mycoplasmas of the genera, Mycoplasma in all morphological aspects (lack cell wall bounded by unit membrane have cytoplasm ribosomes and strands of nuclear material. Their shape is usually spherical to avoid or irregularly tubular to filamentous and their sizes comparable to those of the typical Mycoplasma.

a. Plant Mycoplasma like organism are generally present in the sap of a small number of phloem sieve tubes.
b. Most plant Mycoplasma like organisms are transmitted from plant to plant by leaf hopper but some are transmitted by psyllids and plant hoppers.
c. Plant Mycoplasma like bodies also grow in the alimentary panals, hemolymph salivary glands and intracellularly in the various body organs of their insect vectors.
d. The vector cannot transmit the Mycoplasma, immediately after feeding on infected plants, but it begins to transmit the Mycoplasma, immediately after feeding on depending upon temperature. Shorter incubation period occurs at 30 0 C where as longest at about 10 0C.

The incubation period is required for multiplication and distribution of the Mycoplasms within the insect. If Mycoplasma is acquired from the plant, it multiplies first in the intestinal cells of vector, it then passes into the hemolymph and internal organs are infected. When the concentration of mycoplasms in salivary glands reachs to a certain level, the insect begins to transmit the pathogen to new plant and continuous to do so for the rest of its life.

MLOs are usually be acquired readily or better by nymph than adult hopper and survive through subsequent molts, but are not passed on the adult to the egg and to next generation.

Symptoms Produced by Mycoplasma Like Organisms (MLOs) in Plants

1. Mycoplasma like bodies are now stated to be occur in more than 60 to 70 plant diseases, which are characterised by the growth abnormalities and yellowing of leaves.
2. Characteristic symptom of yellow type disease includes uniform yellowing or reddening of leaves, smaller leaves, shortening of internodes, stunting of plants and proliferation of auxiliary bud.
3. "Witch brooms" includes reduction of leaf size with leaves becoming brittle, excessive proliferation of shoots.

4. "Phyllody" replacement of floral parts of leaves greening or sterility of flowers and reduced yield. Finally more or less rapid dieback, decline and disorder after several years.
5. MLOs are mostly restricted to phloem tissue becomes; a phloem may provide favourable conditions for growth as phloem elements have a high osmotic pressure and slightly alkaline PH.

Transmission of Mycoplasms Like Organisms (MLOs)

1. MLOs are found exclusively in phloem tissue of infected plants and require a vector for pheir dispersal.
2. A majority of MLOs are transmitted by leaf hopper, tree hoppers, plant hopper, aphids, and mites also play a role in transmission of MLOs.
3. The vector acquires the MLOs while feeding on contaminated phloem. Incubation period ranges between 10 to 45 days depending upon the temperature. During this period MLOs multiply in the body organs of vectors particularly in salivary glands and hemolymph.
4. MLOs are not passed on through the eggs of contaminated insects.
5. In addition to vectors, MLOs are also transmitted by grafting and by the plant parasite like dodder.

Disease Caused by Mycoplasma Like Organisms (MLOs)

1. Aster yellow
2. Lethal yellowing of coconut
3. Elm phloem necrosis (Elm yellow)
4. Peach (X-Disease)
5. Pear decline
6. Citrus stubborur
7. Corn stunt
8. Sandle spike
9. Citrus greening
10. Seasamum phyllody
11. Grassy shoot of sugarcane
12. Little leaf of Brinjal
13. Rice yellow dwarf.

General Control Measures of Mycoplasma Like Organisms

1. Most of the control measures against yellow type of disease are avoided at preventing infections.
2. Use of antibiotics: Antibiotics are known to suppress the yellow type disease. The most effective antibiotics appear to be chloro tetracycling, oxy tetra cycline and tetra cycline with choramphenicol, ledermycin and metha cycline. The application of antibiotics by root dip or paste under tha bark, standing cutting in solution appears to be more effective than foliar spray or soil drenches.
3. Heat Treatment: It is possible to cure yellow infected plants grassy, shoot of sugarcane by heat treatment.
4. Vector Control: Since most of disease are spread by the vector, control of vector through effective insecticides like rogar, Malathion, Endosulphan, demecron etc.

Comparison between Mycoplasma and Virus

Sr.No	*Mycoplasma*	*Virus*
1	Occurs in nature	Occurs naturally
2	Lack cell wall or cell nucleopeptide	No cell wall or membrane only naked nucleo protecteins.
3	Can be filtered through bacteria proof filter.	Filter through bacteria proof filters.
4	Mostly free living	Obligate parasites.
5	Mostly require sterol for growth.	Not required
6	Limited metabolic activities	Lack enzymes grow at the expense of other host.
7	RNA and DNA both are present in cell.	Either RNA or DNA.
8	Size 0.1 to 1 micron	Variable
9	Shape polymorphic,pleomorphic	Rigid
10	Synthesize protein by their own enzyme	No
11	Not dependent on host nucleic acid for multiplication.	Dependent

Definition: Plant disease in which no foreign organisms or parasitic is associate with the cause is known as Non-parasitic disease. They differ from virus disease in being no infectious.

General Characteristics

1. Non infectious disease of plants is caused by the lack of excess of something that supports life.
2. Non infectious disease occurs in the absence of pathogen and therefore, cannot be transmitted from diseased to healthy plants.
3. Non infectious disease may effect plant in all stages of their lives, such as seeds, seedlings, mature plants or fruits.
4. These disease cause damage in field in storage or at the market.
5. Symptoms may range from slight to severe and affected plants may even die.

Factors Responsible for Non Parasitic Disease

1. Unfavourable Temperature:

   a. High Temperature Effects:

1. Plants are generally injured faster and to a greater extent when temperature becomes higher than maximum for growth.
2. High temperature are usually responsible for sun solid injuries appearing the sun exposed sides of fleshy fruits and vegetables.

Ex. Sun scald of apple, peppers, tomato, onion- bulbs, potato tubers, canker of linseeds.

   b. Low Temperature Effects:

1. Low temperature causes greater damage to crops than high temperature.
2. Blotch type necrosis in Potato is due to freezing injury.
3. The frost injuries involved killing of buds of peach, cherry, killing of flowers, young fruits and sometimes succulent twings of most trees.
4. Low winter temperature may kill young roots of trees such as apple and may also cause bark splitting and canker development.

Unfavourable Light: Lack of sufficient light or absence of light retards chlorophyll formation and promotes slender growths with long internodes, thus leading to pale green leaves, spridly growth and premature drop of leaves and flower. This condition is known as "Etiloation".

Moisture Effects

Low Moisture Effects: Plant suffering from lack of sufficient soil moisture usually remain stunted are pale green to light yellow, have few small dropping leaves and finally in absence of moisture plant dry and wilt.

High Moisture Effects: Stagnation of water or excessive moisture by flooding in the field may cause decay or rotting of fibrous root of plants resulting into wilting. Primarily due to accumulation of toxic materials around the root and base of stem and also due to non availability of nutrients.

Ex. Tip burn of paddy.

Another common symptom of house plants caused by excessive moisture is swelling. Edema appears as small bumps on the lower of leaves or on stem.

Inadequate Oxygen: When there is excessive respiration in closed atmosphere the entire oxygen supply may be exhausted resulting in disintegration of cells due to enzymic action.

Ex. Black heart of potato.

Atmospheric Impurities or Air Pollution: Presence of injuries gases in the atmosphere may cause definite injury to plant and plant parts.

More serious and wide spread damage is caused to plant in the fields by chemicals such as Ozone, sulphur dioxide hydrogen fluoride, nitrogen dioxide, peroxyaacotyl nitrates.

Ex. Black tip of mango necrosis. Fruit on trees is close proximity to brick kilns may bear necrotic lesions and become useless for sale and consumption. The smoke of kills polluted the air with toxic gases like gases like sulphur dioxide which caused necrosis of tissues.

Toxic effects of Decomposition Organic Matter in Soil and Soil Mineral Toxicity:

Crop residue decomposing in soil produce toxic substances such as fatty acids which produce symptoms of damping off, root rot, wilt and nutritional deficiency.

Excessive amount of sodium salt especially sodium chloride, sodium sulphate and sodium carbonate raise the PH of the soil and cause alkali injury i.e Chlorosis, stunting, boron, manganese and copper have been most frequently implicated in mineral toxicity disease. Excess boron is toxic to many vegetable and trees, Excess manganese is known to cause crinkle leaf disease in cotton.

Herbicidal Injury: Some of the most frequent plant disorder seems to be the result of extensive use of herbicides. Increasing number of herbicides in use for general or specific weed control has created problems.

Nutritional Deficiencies or Disorder in Plants:

i) Deficiency of minerals viz Nitrogen, Phosphorus, potash, manganese , magnesium, boron, zinc, copper , iron, etc. results in disorders in plant metabolism and cause hunger signs in the crops.

ii) Excess of minerals disturbs nutritional balance needed for good metabolism in the palnt, thus hindering the effect of essential element.

iii) The deficiencies and excess of minerals also reduce the resistance of plant to fungal, bacterial and other diseases.

iv) The kind of symptoms produced by deficiency of a certain nutrient depend primarily on the functions of that particular element in the plant.

Plant Disease Due to Lack of Minerals

Other Improper Agricultural Practises: A variety of other agricultural practices improperly carried out, many cause consideration damage to plants. Improper application of chemical viz. fungicides insecticides, nematicides and fertilizers at too high concentration results, into losses. Spray injury results into leaf burn or spolting or rosseting of fruits. Excessive or deep cultivation between rows of growing plants may pull roots of many plants.

Flowering Plant of Parasites and Its Classification

1. Although most of the plant disease are caused by fungi, bacteria, viruses and nematodes. There are few seed plants, which are parasitic on living plants.
2. In many cases the damage caused by these parasites is slight on attacked host plants are little importance, but there are some examples where those flowering plants attack valuable crops and trees causing considerable losses.
3. Some of these attack on root while others parasitize stem, some are devoid of chlorophyll and entirely depended on their hosts for food supply while other have chlorophyll and obtain only mineral constituents of food from the host.
4. More than 2500 species of higher plants are known to live parasitically on other plants.

Classification of Flowering Plant Parasites

i) Complete Parasite (Holoparasites):
   a. Root Orobanchea (Broom rape)
   b. Stem Cuscuta (Dodder Amarvel)

ii) Partial Parasites (Semi Parasites)
   a. Root Striga (Sandle wood witch weed)
   b. Stem Loranthus (Banda or Deudrophae)

Complete Stem Parasite – Dodder

Family: Cuscutaceae

Genus: Cuscuta

1. These are non-chlorophyll bearing leafless, twining parasitic seed plants. They are yellow, pink or orange in colour, they attach to the host.
2. The first appearance of parasite in field is noticed as small masses of branched, thread like, leafless stem, which are devoid of green pigment and twice around the stem or leaves of the hosts.
3. The leaves are represented by minutes functionless scales, which are evident on close examination, when the stem of parasites comes in contact with the hosts, the minute root like organs, haustoria penetrates into the host cortex and serves as an organ of food absorption.
4. When the relationship with the host is firmly established, the dodder plant losses the contact from soil.
5. The flowers are found in clusters, they are tiny, white, pink or yellowing in colour. The seeds are formed in capsule. A single plant may produce as many as 3000 seeds.
6. Dodder may also serves as a bridge for transmission of viruses from virus infected to virus free plant. There are 3 species of dodder:
   i) Larger seed: - Cuscuta indicora
   ii) Small seed: - Cuscuta planiflora
   iii) Field dodder:- Cuscuta campestris
7. The common dodder Cuscuta granovil attacks clovers, berseem flag and many other oilseed crops. It also attacks ornamental and hedge plants. It is fast developing parasites and within 2-3 seasons may destroy a complete plant.

8. The dodder perepetuates through seeds which remain dormant in the soil unitl favourable seasons returns, stem portion of parasite is also meant of perpetuation.

Control Measures of Dodder:

1. Selection of dodder free seeds for sowing.
2. Preventing the movement of grazing animals from infected field to clean field.
3. Restriction of flow of irrigation water through infected field.
4. Destruction of flow of irrigation water through infected field.
5. Crop rotation with non host plants.
6. Dodder can be controlled by the use of soil herbicides, such as chloroproopham, DCPA, Dichobhil, Donoseb, glyphosate, these chemicals kills the dodder plant upon its germiantion from seed.

Complete Root Parasite - Broom Rape

Family: Orobanchaceae

Genus: Orobanche.

1. In some areas of the world, broom rape cause losses varying from 15 to 70 % of the crop, Orobanche spp. are total root parasites affecting tobacco, Brinjal, Tomato, Cauliflower, Turnip and many other solanaceous and cruciferous plants.
2. It is especially destructive to tobacco and Brinjal. The parasite consists of a stount, fleshy stem 15 to 50 cm long. This stem is yellow or brownish red in colour and is covered by small thin and brown scaly leaves.
3. The flower appearing in the axil of the leaves are white and tubular seeds are very small and black in colour and many remain viable in soil for several years.
4. The haustoria of the parasite penetrate into the root of the hosts and draw it nourishment. When the host is carefully uprooted, the parasitic roots are seen intertwined with the host root systems; the growth of the host is retarded and remains stunted.

Control Measures of Broom Rape:

1. Destroy the parasite before flowering.
2. Long crop rotation.
3. Spraying the soil with 0.25% copper sulphate (drenching) solution has been reported to be successful in destroying the parasites.

4. Fumigating the soil with methyl bromide.
5. It was recently reported that broom tapes were effectively control after treatment with the herbicides "Glyphosate".

Partial Stem Parasite - Loranthus or Bandgul

1. It is common parasites of mango trees. In Northern India 60-90% of mango trees and large no of other trees are heavily or moderately infected by these parasites.
2. Loranthus Dendro-phthae falcate, the most common species in India is the semi parasitic of the tree trunk and branches. Their leaves posses chlorophyll and synthesize carbohydrate constituent of their food requirement.
3. Since the parasite attacks the aerial part of host tree, situated far above the soil level and as such devoid of root system of its own.
4. The parasite attacks, the aerial parts of the host trees, by developing haustoria and obtain its nourishment directly from the vascular system of the host plant.
5. The continuous sucking of the food material by parasite resulting the host to die.
6. The place at which host is attached and where the haustoria penetrates often swell to form tumours which vary in size according to age of the parasites.
7. The flowers of parasites are borne in clusters. They are long tabular in shape and usually greenish, red or white in colour according to species.
8. The fruit is fleshy and contains a solitary seed. It is sweet eaten by birds and animals.
9. The parasite is spread by dispersal of its seed, mostly through birds and to some extent by other animals.
10. The damage done by the parasite is most marked in production of new growth of the host. The quantity and yield of fruit is considerably lowered. Leaves are reduced in size and show unhealthy green colour.

Control Measures of Loranthus:

1. Scrapping of parasite before seedling from the infected bunches.
2. Sowing of branches sufficient low to the tumours.
3. Injection of CuSo4 or 2-4-D into the infected branches has been found effectively in eradicating the parasite from mango.

Partial Root Parasite – Striga

Family: Scrophulariaceae

Genus: Striga

1. Witch weed (Striga species) is well known partial root parasite of sugarcane Jawar, maize, cereals and millets in India. There are four species of Striga reports in the country on sugarcane, rice , sorghum and other millets.
   1. S.densiflora
   2. S. quphrasioldes
   3. S.lutee
   4. S.asiatica.
2. These plants although obligate parasites do not obtain all of their nutrient material from their host root. They posses' chlorophyll bearing leaves. Plants are 20-60 CM in roots of cereals, sugarcane, etc.
3. Striga can be found on light as well as heavy soil in rabi and kharif season. The seeds of Striga are very minute and produced in great abundance 50,000 to 1, 00,000 seeds /pl/yr.
4. One flower / capsule contains 1200-1500 seeds. Viability of those seeds has been reported to be from 12-40 years. Short distance dissemination of these seeds.
5. For germination of seeds of Striga species, stimulant provided by the root executes of specific host is essential.
6. Seed starts germination after 7-10 days. In most cases germination occurs 40 cm depth. After germination, the parasites grow below the soil surface for about 4-8 weeks and produces underground stem and root.
7. The underground portion of the stem contains bud in the axil of leaf. Stem of parasite forms haustoria which penetrates the root of hosts plants and also water and nutrient eventually wasting and destroying the host.

Striga Difficult to Control Because:

1. Heavy seed production 50,000 to1, 00,000 per plant.
2. Seed viability is longer.
3. Most hyyries are stimulating positive.
4. Alternative host like grasses are present on bunds.

Control Measures of Striga:

1. Complete eradication of parasite before flowering.
2. Regular Interculture should be followed.
3. Crop rotation with cotton-Jowar- Groundnut.
4. Weedicides are used to control Striga before flowering.
   a. 2-4-D @ 2.5 lit/500 lit of H2O per ha.
   b. Attrazine @ 2 kg / 500 lit of water per ha.
   c. 1% TCPA @ 45 kg/ha.
   d. Deep ploughing after harvest reduces the vialibility of seed.

Bioassay

Bioassay (commonly used shorthand for biological assay), or biological standardization is a type of scientific experiment. Bioassays are typically conducted to measure the effects of a substance on a living organism and are essential in the development of new drugs and in monitoring environmental pollutants. Both are procedures by which the potency or the nature of a substance is estimated by studying its effects on living matter. Bioassay is a procedure for the determination of the concentration of a particular constitution of a mixture.

Use

Bioassays are procedures that can determine the concentration of purity or biological activity of a substance such as vitamin, hormone, and plant growth factor. While measuring the effect on an organism, tissue cells, enzymes or the receptor is preparing to be compared to a standard preparation. Bioassays may be qualitative or quantitative. Qualitative bioassays are used for assessing the physical effects of a substance that may not be quantified, such as abnormal development or deformity. An example of a qualitative bioassay includes Arnold Adolph Berthold's famous experiment on castrated chickens. This analysis found that by removing the testes of a chicken, it would not develop into a rooster because the endocrine signals necessary for this process were not available. Quantitative bioassays involve estimation of the concentration or potency of a substance by measurement of the biological response that it produces. Quantitative bioassays are typically analyzed using the methods of biostatistics.

Definition

"The determination of the relative strength of a substance (as a drug) by comparing its effect on a test organism with that of a standard preparation." ....

Purpose

1. Measurement of the pharmacological activity of new or chemically undefined substances
2. Investigation of the function of endogenous mediators
3. Determination of the side-effect profile, including the degree of drug toxicity
4. Measurement of the concentration of known substances (alternatives to the use of whole animals have made this use obsolete)
5. Assessing the amount of pollutants being released by a particular source, such as wastewater or urban runoff.
6. Determining the specificity of certain enzymes to certain substrates.

Types: Bioassays are of two types:

Quantal

A quantal assay involves an "all or none response". For example: Insulin induced hypoglycemic convulsive reaction or the cardiac arrest caused by digitalis. The response is either +ve or -ve, there is no intermediate response e.g.—either convulsion occurs or doesn't occur; similarly is with cardiac arrest.

In case of toxicity studies, the animal receiving a dose of drug either dies or does not die. Also, no intermediate response is possible. This is also known as the "all or none" response assay. The quantal method though not precise is employed for bioassay of substance in the following ways:

(a) Comparison of threshold response or

(b) Comparison of effective dose (ED_{50}) or median lethal dose (LD_{50})

Graded

Graded assays are based on the observation that there is a proportionate increase in the observed response following an increase in the concentration or dose. The parameters employed in such bioassays are based on the nature of the effect the substance is expected to produce. For example: contraction of smooth muscle preparation for assaying histamine or the study of blood pressure response in case of adrenaline.

A graded bioassay can be performed by employing any of the below-mentioned techniques. The choice of procedure depends on:

1. the precision of the assay required
2. the quantity of the sample substance available
3. the availability of the experimental animals.

Techniques

1. Matching Bioassay
2. Interpolation Method
3. Bracketing Method
4. Multiple Point Bioassay (i.e.-Three-point, Four-point and Six Point Bioassay)

Matching Bioassay: It is the simplest type of the bioassay. In this type of bioassay, response of the test substance taken first and the observed response is tried to match with the standard response. Several responses of the standard drug are recorded till a close matching point to that of the test substance is observed. A corresponding concentration is thus calculated. This assay is applied when the sample size is too small. Since the assay does not involve the recording of concentration response curve, the sensitivity of the preparation is not taken into consideration. Therefore, precision and reliability is not very good.

Interpolation bioassay: Bioassays are conducted by determining the amount of preparation of unknown potency required to produce a definite effect on suitable test animals or organs or tissue under standard conditions. This effect is compared with that of a standard. Thus the amount of the test substance required to produce the same biological effect as a given quantity the unit of a standard preparation is compared and the potency of the unknown is expressed as a % of that of the standard by employing a simple formula.

Many times, a reliable result cannot be obtained using this calculation. Therefore it may be necessary to adopt more precise methods of calculating potency based upon observations of relative, but not necessarily equal effects, likewise, statistical methods may also be employed. The data (obtained from either of assay techniques used) on which bioassay are based may be classified as quantal or graded response. Both these depend ultimately on plotting or making assumption concerning the form of DRC.

Environmental Bioassays

Environmental bioassays are generally a broad-range survey of toxicity. A toxicity identification evaluation is conducted to determine what the relevant toxicants are. Although bioassays are beneficial in determining the biological activity within an organism, they can often be time-consuming and laborious. Organism-specific factors may result in data that is not applicable to others in that species. For these reasons, other biological techniques are often employed, including radioimmunoassays.

Water pollution control requirements in the United States require some industrial dischargers and municipal sewage treatment plants to conduct bioassays. These procedures, called whole effluent toxicity tests, include acute toxicity tests as well as chronic test methods. The methods involve exposing living aquatic organisms to samples of wastewater.

Chrysanthemum

Chrysanthemums, sometimes called mums or chrysanths, are flowering plants of the genus *Chrysanthemum* in the family Asteraceae. They are native to Asia and northeastern Europe. Most species originate from East Asia and the centre of diversity is in China. There are about 40 valid species. There are countless horticultural varieties and cultivars.

Etymology

The name "chrysanthemum" is derived from the Greek words *chrysos* (gold) and *anthemon* (flower).

Taxonomy

The genus once included more species, but was split several decades ago into several genera, putting the economically important florist's chrysanthemum in the genus *Dendranthema*. The naming of the genera has been contentious, but a ruling of the International Code of Botanical Nomenclature in 1999 changed the defining species of the genus to *Chrysanthemum indicum*, restoring the florist's chrysanthemum to the genus *Chrysanthemum*.

The other species previously included in the narrow view of the genus *Chrysanthemum* are now transferred to the genus *Glebionis*. The other genera separate from *Chrysanthemum* include *Argyranthemum*, *Leucanthemopsis*, *Leucanthemum*, *Rhodanthemum*, and *Tanacetum*.

Description

Wild *Chrysanthemum* taxa are herbaceous perennial plants or subshrubs. They have alternately arranged leaves divided into leaflets with toothed or occasionally smooth edges. The compound inflorescence is an array of several flower heads, or sometimes a solitary head. The head has a base covered in layers of phyllaries. The simple row of ray florets are white, yellow or red; many horticultural specimens have been bred to bear many rows of ray florets in a great variety of colours. The disc florets of wild taxa are yellow. The fruit is a ribbed achene.

History

Chrysanthemums were first cultivated in China as a flowering herb as far back as the 15th century BC. Over 500 cultivars had been recorded by the year 1630. The plant is renowned as one of the Four Gentlemen in Chinese and East Asian art. The plant is particularly significant during the Double Ninth Festival. The flower may have been brought to Japan in the eighth century AD, and the Emperor adopted the flower as his official seal. The "Festival of Happiness" in Japan celebrates the flower.

Chrysanthemums entered American horticulture in 1798 when Colonel John Stevens imported a cultivated variety known as 'Dark Purple' from England. The introduction was part of an effort to grow attractions within Elysian Fields in Hoboken, New Jersey.

Economic Uses

Ornamental uses:

***Figure:** 'Dance'*

Figure: *'Enbee Wedding Golden' and 'Feeling Green'*

Modern cultivated chrysanthemums are showier than their wild relatives. The flower heads occur in various forms, and can be daisy-like or decorative, like pompons or buttons. This genus contains many hybrids and thousands of cultivars developed for horticultural purposes. In addition to the traditional yellow, other colours are available, such as white, purple, and red. The most important hybrid is *Chrysanthemum* × *morifolium* (syn. *C.* × *grandiflorum*), derived primarily from *C. indicum*, but also involving other species.

Over 140 varieties of chrysanthemum have gained the Royal Horticultural Society's Award of Garden Merit.

Chrysanthemums are divided into two basic groups, garden hardy and exhibition. Garden hardy mums are new perennials capable of wintering in most northern latitudes. Exhibition varieties are not usually as sturdy. Garden hardies are defined by their ability to produce an abundance of small blooms with little if any mechanical assistance, such as staking, and withstanding wind and rain. Exhibition varieties, though, require staking, overwintering in a relatively dry, cool environment, and sometimes the addition of night lights.

The exhibition varieties can be used to create many amazing plant forms, such as large disbudded blooms, spray forms, and many artistically trained forms, such as thousand-bloom, standard (trees), fans, hanging baskets, topiary, bonsai, and cascades.

Chrysanthemum blooms are divided into 13 different bloom forms by the US National Chrysanthemum Society, Inc., which is in keeping with the international classification system. The bloom forms are defined by the way in which the ray and disk florets are arranged. Chrysanthemum blooms are composed of many individual flowers (florets), each one capable of producing a seed. The disk florets are in the centre of the bloom head, and the ray florets are on the perimeter. The ray florets are considered imperfect flowers, as they only possess the female productive organs, while the disk florets are considered perfect flowers, as they possess both male and female reproductive organs.

Irregular incurves are bred to produce a giant head called an *ogiku*. The disk florets are concealed in layers of curving ray florets that hang down to create a 'skirt'. Regular incurves are similar, but usually with smaller blooms and a dense, globular form. Intermediate incurve blooms may have broader florets and a less densely flowered head.

In the reflex form, the disk florets are concealed and the ray florets reflex outwards to create a mop-like appearance. The decorative form is similar to reflex blooms, but the ray florets usually do not radiate at more than a 90° angle to the stem.

The pompon form is fully double, of small size, and very globular in form. Single and semidouble blooms have exposed disk florets and one to seven rows of ray florets.

In the anemone form, the disk florets are prominent, often raised and overshadowing the ray florets. The spoon-form disk florets are visible and the long, tubular ray florets are spatulate.

In the spider form, the disk florets are concealed, and the ray florets are tube-like with hooked or barbed ends, hanging loosely around the stem. In the brush and thistle variety, the disk florets may be visible.

Culinary Uses

Yellow or white chrysanthemum flowers of the species *C. morifolium* are boiled to make a sweet drink in some parts of Asia. The resulting beverage is known simply as chrysanthemum tea. In Korea, a rice wine flavored with chrysanthemum flowers is called *gukhwaju*.

Chrysanthemum leaves are steamed or boiled and used as greens, especially in Chinese cuisine. The flowers may be added to thick snakemeat soup to enhance the aroma. Small chrysanthemums are used in Japan as a sashimi garnish.

Insecticidal Uses

Pyrethrum (*Chrysanthemum* [or *Tanacetum*] *cinerariaefolium*) is economically important as a natural source of insecticide. The flowers are pulverized, and the active components, called pyrethrins, which occur in the achenes, are extracted and sold in the form of an oleoresin. This is applied as a suspension in water or oil, or as a powder. Pyrethrins attack the nervous systems of all insects, and inhibit female mosquitoes from biting. In sublethal doses they have an insect repellent effect. They are harmful to fish, but are far less toxic to mammals and birds than many synthetic insecticides. They are not persistent, being biodegradable, and also decompose easily on exposure to light. Pyrethroids such as permethrin are synthetic insecticides based on natural pyrethrum.

Environmental Uses

Chrysanthemum plants have been shown to reduce indoor air pollution by the NASA Clean Air Study.

Cultural Significance and Symbolism

In some countries of Europe (e.g., France, Belgium, Italy, Spain, Poland, Hungary, Croatia), incurve chrysanthemums are symbolic of death and are used only for funerals or on graves, while other types carry no such symbolism; similarly, in China, Japan and Korea, white chrysanthemums are symbolic of lamentation and/or grief. In some other countries, they represent honesty. In the United States, the flower is usually regarded as positive and cheerful, with New Orleans as a notable exception.

Australia

- In Australia,the chrysanthemum is sometimes given to mothers for Mother's Day, which falls in May in the southern hemisphere's autumn. Men may sometimes also wear it in their lapels to honour mothers, although the flower is naturally in season during the southern hemisphere's spring and early summer.

China

- The chrysanthemum is one of the "Four Gentlemen" of China (the others being the plum blossom, the orchid, and bamboo). The chrysanthemum is said to have been favoured by Tao Qian, an influential Chinese poet, and is symbolic of nobility. It is also one of the four symbolic seasonal flowers.

- A chrysanthemum festival is held each year in Tongxiang, near Hangzhou, China.
- Chrysanthemums are the topic in hundreds of poems of China.
- The "golden flower" referred to in the 2006 movie *Curse of the Golden Flower* is a chrysanthemum.
- "Chrysanthemum Gate", often abbreviated as Chrysanthemum, is taboo slang meaning "anus" (with sexual connotations).
- Chrysanthemums were first cultivated in China as a flowering herb as far back as the 15th century BC.
- An ancient Chinese city (Xiaolan Town of Zhongshan City) was named Ju-Xian, meaning "chrysanthemum city".

Germany

- Industrial musicians Einstürzende Neubauten base their song "Blume" around the flower.

Japan

- The Chrysanthemum Throne is the name given to the position of Japanese emperor.
- Chrysanthemum crest is a general term for a mon of chrysanthemum blossom design; there are more than 150 patterns. The Imperial Seal of Japan is a particularly notable one; it is used by members of the Japanese imperial family. There are also a number of formerly state-endowed shrines which have adopted a chrysanthemum crest, most notably Tokyo's Yasukuni Shrine.
- The Supreme Order of the Chrysanthemum is a Japanese honour awarded by the emperor.
- The city of Nihonmatsu, Japan hosts the "Nihonmatsu Chrysanthemum Dolls Exhibition" every autumn in historical ruin of Nihonmatsu Castle.
- In Imperial Japan, small arms were required to be stamped with the Imperial Chrysanthemum, as they were considered the personal property of the Emperor.
- The chrysanthemum is also considered to be the seasonal flower of September.

United States

- The chrysanthemum was recognized as the official flower of the city of Chicago by Mayor Richard J. Daley in 1966.

- The chrysanthemums is the official flower of the city of Salinas, California.
- The yellow chrysanthemum is the official flower of the sorority Sigma Alpha and the pharmacy fraternity Lambda Kappa Sigma .
- The white chrysanthemum is the official flower of Triangle Fraternity.

Others

- The term “chrysanthemum” is also used to refer to a certain type of fireworks shells that produce a pattern of trailing sparks similar to a chrysanthemum flower.
- The chrysanthemum is also the flower of November.

Tutankhamen was buried with floral collars of chrysanthemum. Resins of the plant were used in incense cones used to ward off insects.

Aster (Genus)

Aster is a genus of flowering plants in the family Asteraceae. Its circumscription has been narrowed, and it now encompasses around 180 species, all but one of which are restricted to Eurasia; many species formerly in *Aster* are now in other genera of the tribe Astereae.

Circumscription

The genus *Aster* once contained nearly 600 species in Eurasia and North America, but after morphologic and molecular research on the genus during the 1990s, it was decided that the North American species are better treated in a series of other related genera. After this split there are roughly 180 species within the genus, all but one being confined to Eurasia. The name *Aster* comes from the Ancient Greek word *astér*, meaning “star”, referring to the shape of the flower head. Many species and a variety of hybrids and varieties are popular as garden plants because of their attractive and colourful flowers. *Aster* species are used as food plants by the larvae of a number of Lepidoptera species. Asters can grow in all hardiness zones.

The genus *Aster* is now generally restricted to the Old World species, with *Aster amellus* being the type species of the genus, as well as of the family Asteraceae. The New World species have now been reclassified in the genera *Almutaster*, *Canadanthus*, *Doellingeria*, *Eucephalus*, *Eurybia*, *Ionactis*, *Oligoneuron*, *Oreostemma*, *Sericocarpus* and *Symphyotrichum*, though all are treated within the tribe Astereae. Regardless of the taxonomic change, all are still widely referred to as "asters" (popularly "Michaelmas daisies" because of their typical blooming period) in the horticultural trades.

Some common North American species that have now been moved are:

- *Aster breweri* (now *Eucephalus breweri*), Brewer's aster
- *Aster cordifolius* (now *Symphyotrichum cordifolium*), blue wood aster
- *Aster dumosus* (now *Symphyotrichum dumosum*), New York aster
- *Aster divaricatus* (now *Eurybia divaricata*), white wood aster
- *Aster ericoides* (now *Symphyotrichum ericoides*), heath aster
- *Aster laevis* (now *Symphyotrichum laeve*), smooth aster
- *Aster lateriflorus* (now *Symphyotrichum lateriflorum*), "Lady in Black", calico aster
- *Aster novae-angliae* (now *Symphyotrichum novae-angliae*), New England aster
- *Aster novi-belgii* (now *Symphyotrichum novi-belgii*), New York aster
- *Aster peirsonii* (now *Oreostemma peirsonii*), Peirson's aster
- *Aster protoflorian* (now *Symphyotrichum pilosum)*, frost aster
- *Aster scopulorum* (now *Ionactis alpina*), lava aster
- *Aster sibiricus* (now *Eurybia sibirica*), Siberian aster

The "China aster" is in the related genus *Callistephus*.

Species

In the United Kingdom, there are only two native members of the genus: goldilocks, which is very rare, and *Aster tripolium*, the sea aster. *Aster alpinus* spp. *vierhapperi* is the only species native to North America.

Some common species are:

- *Aster alpinus*, Alpine aster
- *Aster amellus*, European Michaelmas daisy or Italian aster
- *Aster linosyris*, goldilocks aster
- *Aster scaber*
- *Aster tataricus*, Tatarian aster
- *Aster tongolensis*
- *Aster tripolium*, sea aster

Hybrids and Cultivars

(those marked AGM have gained the Royal Horticultural Society's Award of Garden Merit:-

- *Aster* × *frikartii* (*A. amellus* × *A. thomsonii*) Frikart's aster
    - o *Aster* × *frikartii* 'Mönch'AGM
    - o *A.* × *frikartii* 'Wunder von Stäfa'AGM
- 'Kylie' (*A. novae-angliae* 'Andenken an Alma Pötschke' × *A. ericoides* 'White heather')
- 'Ochtendgloren'AGM (*A. pringlei* hybrid)
- 'Photograph'AGM

6

Soil Contamination

Soil contamination or soil pollution is caused by the presence of xenobiotic (human-made) chemicals or other alteration in the natural soil environment. It is typically caused by industrial activity, agricultural chemicals, or improper disposal of waste. The most common chemicals involved are petroleum hydrocarbons, polynuclear aromatic hydrocarbons (such as naphthalene and benzo(a)pyrene), solvents, pesticides, lead, and other heavy metals. Contamination is correlated with the degree of industrialization and intensity of chemical usage.

The concern over soil contamination stems primarily from health risks, from direct contact with the contaminated soil, vapors from the contaminants, and from secondary contamination of water supplies within and underlying the soil. Mapping of contaminated soil sites and the resulting cleanup are time consuming and expensive tasks, requiring extensive amounts of geology, hydrology, chemistry, computer modelling skills, and GIS in Environmental Contamination, as well as an appreciation of the history of industrial chemistry.

In North America and Western Europe that the extent of contaminated land is best known, with many of countries in these areas having a legal framework to identify and deal with this environmental problem. Developing countries tend to be less tightly regulated despite some of them having undergone significant industrialization.

Causes

Soil contamination can be caused by:

- Application of pesticides and fertilizers
- Mining

- Oil and fuel dumping
- Disposal of coal ash
- Leaching from landfills
- Drainage of contaminated surface water into the soil
- Discharging urine and faeces in the open

The most common chemicals involved are petroleum hydrocarbons, solvents, pesticides, lead, and other heavy metals.

Coal ash

Historical deposition of coal ash used for residential, commercial, and industrial heating, as well as for industrial processes such as ore smelting, were a common source of contamination in areas that were industrialized before about 1960. Coal naturally concentrates lead and zinc during its formation, as well as other heavy metals to a lesser degree. When the coal is burned, most of these metals become concentrated in the ash (the principal exception being mercury). Coal ash and slag may contain sufficient lead to qualify as a "characteristic hazardous waste", defined in the USA as containing more than 5 mg/L of extractable lead using the TCLP procedure. In addition to lead, coal ash typically contains variable but significant concentrations of polynuclear aromatic hydrocarbons (PAHs; e.g., benzo(a)anthracene, benzo(b)fluoranthene, benzo(k)fluoranthene, benzo(a)pyrene, indeno(cd)pyrene, phenanthrene, anthracene, and others). These PAHs are known human carcinogens and the acceptable concentrations of them in soil are typically around 1 mg/kg. Coal ash and slag can be recognized by the presence of off-white grains in soil, gray heterogeneous soil, or (coal slag) bubbly, vesicular pebble-sized grains.jk

Sewage

Treated sewage sludge, known in the industry as biosolids, has become controversial as a fertilizer to the land. As it is the byproduct of sewage treatment, it generally contains more contaminants such as organisms, pesticides, and heavy metals than other soil.

In the European Union, the Urban Waste Water Treatment Directive allows sewage sludge to be sprayed onto land. The volume is expected to double to 185,000 tons of dry solids in 2005. This has good agricultural properties due to the high nitrogen and phosphate content. In 1990/1991, 13% wet weight was sprayed onto 0.13% of the land; however, this is expected to rise 15 fold by 2005. Advocates say there is a need to control this so that pathogenic microorganisms do not get into water courses and to ensure that there is no accumulation of heavy metals in the top soil.

Pesticides and Herbicides

A pesticide is a substance or mixture of substances used to kill a pest. A pesticide may be a chemical substance, biological agent (such as a virus or bacteria), antimicrobial, disinfectant or device used against any pest. Pests include insects, plant pathogens, weeds, mollusks, birds, mammals, fish, nematodes (roundworms) and microbes that compete with humans for food, destroy property, spread or are a vector for disease or cause a nuisance. Although there are benefits to the use of pesticides, there are also drawbacks, such as potential toxicity to humans and other organisms.

Herbicides are used to kill weeds, especially on pavements and railways. They are similar to auxins and most are biodegradable by soil bacteria. However, one group derived from trinitrotoluene (2:4 D and 2:4:5 T) have the impurity dioxin, which is very toxic and causes fatality even in low concentrations. Another herbicide is Paraquat. It is highly toxic but it rapidly degrades in soil due to the action of bacteria and does not kill soil fauna.

Insecticides are used to rid farms of pests which damage crops. The insects damage not only standing crops but also stored ones and in the tropics it is reckoned that one third of the total production is lost during food storage. As with fungicides, the first insecticides used in the nineteenth century were inorganic e.g. Paris Green and other compounds of arsenic. Nicotine has also been used since the late eighteenth century.

There are now two main groups of synthetic insecticides -

Organochlorines include DDT, Aldrin, Dieldrin and BHC. They are cheap to produce, potent and persistent. DDT was used on a massive scale from the 1930s, with a peak of 72,000 tonnes used 1970. Then usage fell as the harmful environmental effects were realized. It was found worldwide in fish and birds and was even discovered in the snow in the Antarctic. It is only slightly soluble in water but is very soluble in the bloodstream. It affects the nervous and endocrine systems and causes the eggshells of birds to lack calcium causing them to be easily breakable. It is thought to be responsible for the decline of the numbers of birds of prey like ospreys and peregrine falcons in the 1950s - they are now recovering.

As well as increased concentration via the food chain, it is known to enter via permeable membranes, so fish get it through their gills.

As it has low water solubility, it tends to stay at the water surface, so organisms that live there are most affected. DDT found in fish that formed part of the human food chain caused concern, but the levels found in the liver, kidney and brain tissues was less than 1 ppm and in fat was 10 ppm which was below the level likely to cause harm. However, DDT was banned in the UK and the United States to stop the further build up of it in the food chain. U.S. manufactureres continued to sell DDT to developing countries, who could not afford the expensive replacement chemicals and who did not have such stringent regulations governing the use of pesticides.

Health Effects

Contaminated or polluted soil directly affects human health through direct contact with soil or via inhalation of soil contaminants which have vaporized; potentially greater threats are posed by the infiltration of soil contamination into groundwater aquifers used for human consumption, sometimes in areas apparently far removed from any apparent source of above ground contamination.

Health consequences from exposure to soil contamination vary greatly depending on pollutant type, pathway of attack and vulnerability of the exposed population. Chronic exposure to chromium, lead and other metals, petroleum, solvents, and many pesticide and herbicide formulations can be carcinogenic, can cause congenital disorders, or can cause other chronic health conditions. Industrial or man-made concentrations of naturally occurring substances, such as nitrate and ammonia associated with livestock manure from agricultural operations, have also been identified as health hazards in soil and groundwater.

Chronic exposure to benzene at sufficient concentrations is known to be associated with higher incidence of leukemia. Mercury and cyclodienes are known to induce higher incidences of kidney damage, some irreversible. PCBs and cyclodienes are linked to liver toxicity. Organophosphates and carbomates can induce a chain of responses leading to neuromuscular blockage. Many chlorinated solvents induce liver changes, kidney changes and depression of the central nervous system. There is an entire spectrum of further health effects such as headache, nausea, fatigue, eye irritation and skin rash for the above cited and other chemicals. At sufficient dosages a large number of soil contaminants can cause death by exposure via direct contact, inhalation or ingestion of contaminants in groundwater contaminated through soil.

The Scottish Government has commissioned the Institute of Occupational Medicine to undertake a review of methods to assess risk to human health from contaminated land. The overall aim of the project is to work up guidance that should be useful to Scottish Local Authorities in assessing whether sites represent a significant possibility of significant harm (SPOSH) to human health. It is envisaged that the output of the project will be a short document providing high level guidance on health risk assessment with reference to existing published guidance and methodologies that have been identified as being particularly relevant and helpful. The project will examine how policy guidelines have been developed for determining the acceptability of risks to human health and propose an approach for assessing what constitutes unacceptable risk in line with the criteria for SPOSH as defined in the legislation and the Scottish Statutory Guidance.

Ecosystem Effects

Not unexpectedly, soil contaminants can have significant deleterious consequences for ecosystems. There are radical soil chemistry changes which can arise from the presence of many hazardous chemicals even at low concentration of the contaminant species. These changes can manifest in the alteration of metabolism of endemic microorganisms and arthropods resident in a given soil environment. The result can be virtual eradication of some of the primary food chain, which in turn could have major consequences for predator or consumer species. Even if the chemical effect on lower life forms is small, the lower pyramid levels of the food chain may ingest alien chemicals, which normally become more concentrated for each consuming rung of the food chain. Many of these effects are now well known, such as the concentration of persistent DDT materials for avian consumers, leading to weakening of egg shells, increased chick mortality and potential extinction of species.

Effects occur to agricultural lands which have certain types of soil contamination. Contaminants typically alter plant metabolism, often causing a reduction in crop yields. This has a secondary effect upon soil conservation, since the languishing crops cannot shield the Earth's soil from erosion. Some of these chemical contaminants have long half-lives and in other cases derivative chemicals are formed from decay of primary soil contaminants.

Cleanup Options

Clean up or environmental remediation is analyzed by environmental scientists who utilize field measurement of soil chemicals and also apply computer models (GIS in Environmental Contamination) for analyzing transport and fate of soil chemicals. There are several principal strategies for remediation:

- Excavate soil and take it to a disposal site away from ready pathways for human or sensitive ecosystem contact. This technique also applies to dredging of bay muds containing toxins.
- Aeration of soils at the contaminated site (with attendant risk of creating air pollution)
- Thermal remediation by introduction of heat to raise subsurface temperatures sufficiently high to volatize chemical contaminants out of the soil for vapour extraction. Technologies include ISTD, electrical resistance heating (ERH), and ET-DSP^{tm}.
- Bioremediation, involving microbial digestion of certain organic chemicals. Techniques used in bioremediation include landfarming, biostimulation and bioaugmentating soil biota with commercially available microflora.
- Extraction of groundwater or soil vapor with an active electromechanical system, with subsequent stripping of the contaminants from the extract.
- Containment of the soil contaminants (such as by capping or paving over in place).
- Phytoremediation, or using plants (such as willow) to extract heavy metals

By Country

Various national standards for concentrations of particular contaminants include the United States EPA Region 9 Preliminary Remediation Goals (U.S. PRGs), the U.S. EPA Region 3 Risk Based Concentrations (U.S. EPA RBCs) and National Environment Protection Council of Australia Guideline on Investigation Levels in Soil and Groundwater.

People's Republic of China

The immense and sustained growth of the People's Republic of China since the 1970s has exacted a price from the land in increased soil pollution. The State Environmental Protection Administration believes it to be a threat to the environment, to food safety and to sustainable agriculture. According to a scientific sampling, 150 million mi (100,000 square kilometers) of China's cultivated land have been polluted, with contaminated water being used to irrigate a further 32.5 million mi (21,670 square kilometers) and another 2 million mi (1,300 square kilometers) covered or destroyed by solid waste. In total, the area accounts for one-tenth of China's cultivatable land, and is mostly in economically developed areas. An estimated 12 million tonnes of grain are contaminated by heavy metals every year, causing direct losses of 20 billion yuan (US$2.57 billion).

United Kingdom

Generic guidance commonly used in the UK are the Soil Guideline Values published by DEFRA and the Environment Agency. These are screening values that demonstrate the minimal acceptable level of a substance. Above this there can be no assurances in terms of significant risk of harm to human health. These have been derived using the Contaminated Land Exposure Assessment Model (CLEA UK). Certain input parameters such as Health Criteria Values, age and land use are fed into CLEA UK to obtain a probabilistic output.

Guidance by the Inter Departmental Committee for the Redevelopment of Contaminated Land (ICRCL) has been formally withdrawn by the Department for Environment, Food and Rural Affairs (DEFRA), for use as a prescriptive document to determine the potential need for remediation or further assessment.

The CLEA model published by DEFRA and the Environment Agency (EA) in March 2002 sets a framework for the appropriate assessment of risks to human health from contaminated land, as required by Part IIA of the Environmental Protection Act 1990. As part of this framework, generic Soil Guideline Values (SGVs) have currently been derived for ten contaminants to be used as "intervention values". These values should not be considered as remedial targets but values above which further detailed assessment should be considered.

Three sets of CLEA SGVs have been produced for three different land uses, namely

- residential (with and without plant uptake)
- allotments
- commercial/industrial

It is intended that the SGVs replace the former ICRCL values. It should be noted that the CLEA SGVs relate to assessing chronic (long term) risks to human health and do not apply to the protection of ground workers during construction, or other potential receptors such as groundwater, buildings, plants or other ecosystems. The CLEA SGVs are not directly applicable to a site completely covered in hardstanding, as there is no direct exposure route to contaminated soils.

To date, the first ten of fifty-five contaminant SGVs have been published, for the following: arsenic, cadmium, chromium, lead, inorganic mercury, nickel, selenium ethyl benzene, phenol and toluene. Draft SGVs for benzene, naphthalene and xylene have been produced but their publication is on hold. Toxicological data (Tox) has been published for each of these contaminants as well as for benzo[a]pyrene, benzene, dioxins, furans and dioxin-like PCBs, naphthalene, vinyl chloride, 1,1,2,2 tetrachloroethane and 1,1,1,2 tetrachloroethane, 1,1,1 trichloroethane, tetrachloroethene, carbon tetrachloride, 1,2-dichloroethane, trichloroethene and xylene. The SGVs for ethyl benzene, phenol and toluene are dependent on the soil organic matter (SOM) content (which can be calculated from the total organic carbon (TOC) content). As an initial screen the SGVs for 1% SOM are considered to be appropriate.

Weed Control

Weed control is the botanical component of pest control, using physical and chemical methods to stop weeds from reaching a mature stage of growth when they could be harmful to domesticated plants and livestock. In order to reduce weed growth, many "weed control" strategies have been developed in order to contain the growth and spread of weeds.

The most basic is ploughing which cuts the roots of annual weeds. Today, chemical weed killers known as herbicides are widely used.

A plant is often termed weed when it has one or more of the following characteristics:

- Little or no value (as in medicinal, nutritional, or energy)
- Very high growth rate and/or ease of germination
- Exhibits competition to crops, for space, light, water and nutrients

Introduction

Weeds can compete with productive crops or pasture, or convert productive land into unusable scrub. Weeds are also often poisonous, distasteful, produce burrs, thorns or other damaging body parts or otherwise interfere with the use and management of desirable plants by contaminating harvests or excluding livestock.

Weeds tend to thrive at the expense of the more refined edible or ornamental crops. They provide competition for space, nutrients, water and light, although how seriously they will affect a crop depends on a number of factors. Some crops have greater resistance than others- smaller, slower growing seedlings are more likely to be overwhelmed than those that are larger and more vigorous. Onions are one of the crops most susceptible to competition, for they are slow to germinate and produce slender, upright stems. Quick growing, broad leafed weeds therefore have a distinct advantage, and if not removed, the crop is likely to be lost. Broad beans however produce large seedlings, and will suffer far less profound effects of weed competition other than during periods of water shortage at the crucial time when the pods are filling out. Transplanted crops raised in sterile seed or potting compost will have a head start over germinating weed seeds.

Weeds also differ in their competitive abilities, and can vary according to conditions and the time of year. Tall growing vigorous weeds such as fat hen (*Chenopodium album*) can have the most pronounced effects on adjacent crops, although seedlings of fat hen that appear in late summer will only produce small plants. Chickweed (*Stellaria media*), a low growing plant, can happily co-exist with a tall crop during the summer, but plants that have overwintered will grow rapidly in early spring and may swamp crops such as onions or spring greens.

The presence of weeds does not necessarily mean that they are competing with a crop, especially during the early stages of growth when each plant can find the resources it requires without interfering with the others. However, as the seedlings' size increases, their root systems will spread as they each begin to require greater amounts of water and nutrients. Estimates suggest that weed and crop can co-exist harmoniously for around three weeks, therefore it is important that weeds be removed early on in order to prevent competition occurring. Weed competition can have quite dramatic effects on crop growth. Harold A Roberts cites research carried out with onions wherein *"Weeds were carefully removed from separate plots at different times during the growth of the crop and the plots were then kept clean. It was found that after competition had started, the final yield of bulbs was being reduced at a rate equivalent to almost 4% per day. So that by delaying weeding for another fortnight, the yield was cut to less than half that produced on ground kept clean all the time."* (*The Complete Know And Grow Vegetables*, Bleasdale, Salter and others, OUP 1991).

He goes on to record that *"by early June, the weight of weeds per unit area was twenty times that of the crop, and the weeds had already taken from the soil about half of the nitrogen and a third of the potash which had been applied"*.

Perennial weeds with bulbils, such as lesser celandine and oxalis, or with persistent underground stems such as couch grass (*Agropyron repens*) or creeping buttercup (*Ranunculus repens*) are able to store reserves of food, and are thus able to grow faster and with more vigour than their annual counterparts. There is also evidence that the roots of some perennials such as couch grass exude allelopathic chemicals which inhibit the growth of other nearby plants.

Weeds can also host pests and diseases that can spread to cultivated crops. Charlock and Shepherd's purse may carry clubroot, eelworm can be harboured by chickweed, fat hen and shepherd's purse, while the cucumber mosaic virus, which can devastate the cucurbit family, is carried by a range of different weeds including chickweed and groundsel.

However, at times the role of weeds in this respect can be overrated. As far as insect pests are concerned, often the species that live on weeds are not the same as those that attack vegetable crops; *"Tests with the common cruciferous weeds such as shepherds purse have shown that they do not act as hosts for the larvae of the cabbage root fly. One exception was found to be the wild radish, but this is not usually a weed of established vegetable gardens"* (Roberts, *The Complete Know And Grow Vegetables*). However pests such as cutworms may first attack weeds then move on to cultivated crops.

While charlock, a common weed in southeastern USA, may be considered a weed by row crop growers, it is highly valued by beekeepers, who seek out places where it blooms all winter, thus providing pollen for honeybees and other pollinators. Its bloom is resistant to all but a very hard freeze, and even that will only kill it back briefly. By feeding an array of pollinators during a seasonal dearth, it can redound to the farmer's advantage. Many weeds are likewise highly beneficial to pollinators.

Methods

In domestic gardens, methods of weed control include covering an area of ground with several layers of wet newspaper or one black plastic sheet for several weeks. In the case of using wet newspaper, the multiple layers prevent light from reaching all plants beneath, which kills them. Saturating the newspaper with water daily speeds the decomposition of the dead plants. Any weed seeds that start to sprout because of the water will also be deprived of sunlight, be killed, and decompose. After several weeks, all germinating weed seeds present in the ground should be dead. Then the newspaper can be removed and the ground can be planted. The decomposed plants will help fertilize the plants or seeds planted later.

In the case of using the black plastic sheet, the greenhouse effect is used to kill the plants beneath the sheet. A 5–10 cm layer of wood chip mulch on the ground will also prevent most weeds from sprouting. Also, gravel can be spread over the ground as an inorganic mulch. In agriculture, irrigation is sometimes used as a weed control measure such as in the case of paddy fields. Many people find that although the black plastic sheeting is extremely effective at preventing the weeds in areas where it covers, but in actual use it is difficult to achieve full coverage.

Figure: *Weeds are removed manually in large parts of India.*

Knowing how weeds reproduce, spread and survive adverse conditions can help in developing effective control and management strategies. Weeds have a range of techniques that enable them to thrive;

Annual and biennial weeds such as chickweed, annual meadow grass, shepherd's purse, groundsel, fat hen, cleaver, speedwell and hairy bittercress propagate themselves by seeding. Many produce huge numbers of seed several times a season, some all year round. Groundsel can produce 1000 seed, and can continue right through a mild winter, whilst Scentless Mayweed produces over 30,000 seeds per plant. Not all of these will germinate at once, but over several seasons, lying dormant in the soil sometimes for years until exposed to light. Poppy seed can survive 80–100 years, dock 50 or more. There can be many thousands of seeds in a square foot or square metre of ground, thus and soil disturbance will produce a flush of fresh weed seedlings.

"Stale Seed bed" Technique

One technique employed by growers is the 'stale seed bed', which involves cultivating the soil, then leaving it for a week or so.

When the initial flush of weeds has germinated, the grower will lightly hoe off before the desired crop is planted. However, even a freshly cleared bed will be susceptible to airborne seed from elsewhere, as well as seed brought in by passing animals which can carry them on their fur, or from freshly imported manure. The organic solution to the problem of spreading annual weeds lies in regular, properly timed weeding, preferably just before flowering (fortuitously, this is also the time at which they will be of the most value in composting).

This technique is also quite often used by farmers who let weeds germinate then return the soil before crop sowing.

Perennial weeds also propagate by seeding; the airborne seed of the dandelion and the rose-bay willow herb are parachuted far and wide. But they also have an additional range of vegetative means of spreading that gives them their pernicious reputation. Dandelion and dock put down deep tap roots, which, although they do not spread underground, are able to regrow from any remaining piece left in the ground. Removal of the complete tap root is the only sure remedy.

The most persistent of the perennials are those that spread by underground creeping rhizomes that can regrow from the tiniest fragment. These include couch grass, bindweed, ground elder, nettles, rosebay willow herb, Japanese knotweed, horsetail and bracken, as well as creeping thistle, whose tap roots can put out lateral roots. Other perennials put out runners that spread along the soil surface. As they creep along they set down roots, enabling them to colonise bare ground with great rapidity. These include creeping buttercup and ground ivy. Yet another group of perennials propagate by stolons-stems that arch back into the ground to reroot. Most familiar of these is the bramble.

All of the above weeds can be very difficult to eradicate- thick black plastic mulches can be effective to a degree, although will probably need to be left in place for at least two seasons. In addition, hoeing off weed leaves and stems as soon as they appear can eventually weaken and kill the plants, although this will require persistence in the case of plants such as bindweed. Nettle infestations can be tackled by cutting back at least three times a year, repeated over a three-year period. Bramble can be dealt with in a similar way. Some plants are said to produce root exudates that suppress herbaceous weeds. *Tagetes minuata* is claimed to be effective against couch and ground elder, whilst a border of comfrey is also said to act as a barrier against the invasion of some weeds including couch.

Use of Herbicides

The above described methods of weed control avoid using chemicals. They are often used by farmers. However, these methods may damage a fragile soil by restructuring it, hence are not always used. They are those preferred by the organic gardener or organic farmer.

However weed control can also be achieved by the use of herbicides. Selective herbicides kill certain targets while leaving the desired crop relatively unharmed. Some of these act by interfering with the growth of the weed and are often based on plant hormones. Herbicides are generally classified as follows;

- Contact herbicides destroy only that plant tissue in contact with the chemical spray. Generally, these are the fastest acting herbicides. They are ineffective on perennial plants that are able to re-grow from roots or tubers.
- Systemic herbicides are foliar-applied and are translocated through the plant and destroy a greater amount of the plant tissue. Modern herbicides such as glyphosate are designed to leave no harmful residue in the soil.
- Soil-borne herbicides are applied to the soil and are taken up by the roots of the target plant.
- Pre-emergent herbicides are applied to the soil and prevent germination or early growth of weed seeds.

In agriculture large scale and systematic weeding is usually required, often by machines, such as liquid herbicide sprayers, or even by helicopter (such as in the USA), to eliminate the massive amount of weeds present on farming lands.

However there are a number of techniques that the organic farmer can employ such as mulching and carefully timed cutting of weeds before they are able to set seed.

Organic Methods

Typically a combination of methods are used in organic situations.

- Drip irrigation: Rubber hoses and other methods are used to bring water directly to the roots of the desired plants. This limits weed access to water.
- Manually pulling weeds: Labourers are used to pull weeds at various points in the growing process.
- Hot Foam: For example foamstream, the hot water infiltrates the cell walls and the high temperature causes the walls to rupture, the foam acts as a thermal blanket holding the boiling water in place.
- Boiling water: Pour boiling water to weed, they will become more green and then die in few hours. Best for weed in cracks or other hard to reach locations.
- Vinegar: Vinegar kills the visible part of the weed. They will wrinkle and die next day, although the root will still be in place to continue growing.
- Mechanically tilling around plants: Tractors are used to carefully till weeds around the crop plants at various points in the growing process. Besides tilling, other mechanical weed control methods also exist
- Ploughing: Ploughing includes tilling of soil, intercultural ploughing and summer ploughing. Ploughing through tilling of soil uproots the weeds which causes them to die. In summer ploughing is done during deep summers. Summer ploughing also helps in killing pests.
- Crop rotation: Rotating crops with ones that kill weeds by choking them out, such as hemp, *Mucuna pruriens*, and other crops, can be a very effective method of weed control. It is a way to avoid the use of herbicides, and to gain the benefits of crop rotation.
- Weed mat: A weed mat is an artificial mulch, fibrous cloth material, bark or newspaper laid on top of the soil preventing weeds from growing to the surface.

Thermal Methods

There are several thermal methods known to control weed. Hot foam (foamstream) causes the cell walls to rupture, killing the plant. Weed burners heat up soil quickly and destroy superficial parts of the plants. Weed seeds are often heat resistant and even react with an increase of growth on dry heat. Since the 19th century soil steam sterilization is used as a farming technique to clean soil completely from weeds. Several research results confirm the high effectivness of humid heat against weeds and its seeds.

UK Legislation

The *Weeds Act, 1959* is described as *"Preventing the spread of harmful or injurious weeds"*, and is mainly relevant to farmers and other rural settings rather than the allotment or garden-scale grower. There are five 'injurious' (that is, likely to be harmful to agricultural production) weeds covered by the provisions of the Weeds Act. These are:

- Spear thistle (*Cirsium vulgare*)
- Creeping or field thistle (*Cirsium arvense*)
- Curled Dock (*Rumex crispus*)
- Broad leaved dock (*Rumex obtusifolius*)
- Common ragwort (*Senecio jacobaea*) (n.b., this weed is poisonous to livestock. Livestock should not be allowed to graze where ragwort has grown until it is eradicated, and any traces have disintegrated. Ragwort should not be allowed to be harvested in hay or silage for feed).

The Department for Environment, Food and Rural Affairs (DEFRA) provides guidance for the treatment removal of these weeds from infested land. Much of this is oriented towards the use of herbicides, the majority of which may not be acceptable to the organic producer (apart from non-synthetic substances like sulphur, which in some circumstances are accepted within Soil Association standards) but in most cases there are manual techniques that can be used such as digging out the roots, mulching out or carefully timed cutting before seeds are able to spread.

Primary responsibility for weeds control rests with the occupier of the land on which the weeds are growing, therefore it is important to be alert to potential weed problems and to take prompt action. However, it should be remembered that most common farmland weeds are not "injurious" within the meaning of the Weeds Act, and many such plant species have conservation and environmental value. When dealing with complaints under the Weeds Act, DEFRA has a duty in law to try to achieve a reasonable balance among different interests. These include agriculture, countryside conservation and the public in general. Constructive discussion about problems caused by weeds can often result in effective solutions and avoid the need for DEFRA to take official action. In addition to those weeds covered by the 1959 act, under section 14 of the *Wildlife and Countryside Act 1981*, it can be an offence to plant or grow certain specified plants in the wild, including Giant Hogweed and Japanese Knotweed. Problems involving these plants can be referred to the local authority for the area where those weeds are growing as some local authorities have bye-laws controlling these plants. There is no statutory requirement for landowners to remove these plants from their property.

What is Integrated Weed Management

Increased world population will demand more food production, which can only be achieved by increasing crop yields and applying a sustainable approach, i.e. more production with rational use of available resources, which also implies responsible use of land and water and enhanced food diversity. Efforts are needed to reduce crop losses due to pests through the implementation of Integrated Pest Management (IPM) (resistant crop varieties, rational use of pesticides, biocontrol and better cultural practices) without harmful side-effects.

Among the pests, weeds are considered an important biotic constraint to food production. Their competition with crops reduces agricultural output (quantity and quality), and increases external costs by spreading them across farm boundaries. It is also a major constraint to increased farmers' productivity, particularly in developing countries. The labour needed for hand- weeding one hectare of land is not within reach of the small farmer's family.

An integrated weed management approach to land management combines the use of complementary weed control methods such as grazing, herbicide application, land fallowing, and biological control. The resulting combinations provide the best possible solutions to weed problems for land managers. By studying the impact of each of the above methods individually as well as in combination, sustainable management systems can be devised to suit different regions and catchment areas.

Herbicide Application

Herbicides applied at the wrong time of the year can be ineffective. This wastes both time and money for the farmer. Mature plants have already produced large amounts of seeds which simply add to the soil seed bank reserves (CSIRO).

Land Fallowing

The strategic fallowing of land may provide windows of opportunity for perennial pastures to establish. This helps them out-compete weed species. A common characteristic of weedy species is that they are generally first level colonisers. That is, when other vegetation is removed by overgrazing, clearing or ploughing and the ground is left bare, weeds establish quickly and reproduce effectively to maintain that niche. Trials where land is fenced off show that some weed species have difficulty competing with other vegetation (CSIRO).

Biological Control

Biological control agents are generally most effective when established in gullies and rocky knolls of hills. These areas are usually inaccessible and too costly to spray so they can provide a safe haven for agents to retreat to from pastures when there is too much disturbance, such as grazing (CSIRO).

Herbicide Tolerant Crops

The discovery of herbicide-resistant weeds in the early 1970s triggered an interest in mimicking this unintentional development for use in crop breeding. The concomitant progress in molecular genetics made it possible to incorporate resistance genes from unrelated organisms into an otherwise susceptible crop. In other words, we were now able to adapt the biology of the crop to the chemistry of an herbicide, whereas we previously had to adapt chemistry to biology. It must, however, be noted that herbicide-resistant crops (HRCs) were first produced by methods of traditional breeding, whereas the major current HRCs have been produced by genetic engineering, the technology which has unintentionally placed these crops in a fierce debate between those in favour, and those against, the introduction and commercial use of genetically modified (GM) crops.

Ecological Importance of IWM

In most ecosystems, herbicides have become one of the most important components in weed control. There are two reasons to explain the increased use of herbicides, the first being the widespread adoption of high-yielding varieties which created economic incentives for farmers to reduce weed infestation; and the second is the availability of cheap herbicides, indicating that the cost of weed control by herbicides in wet-seeded rice is less than one-fifth of the cost of a single hand-weeding in Illoilo, Philippines (Moody, 1991). Similar situations exist in West Java, Indonesia and the Mekong Delta, Vietnam (Pandy and Pingali, 1996).

Because of the availability of cheap herbicides, it is expected that herbicide usage will continue to increase, both in developed countries, and even in developing countries, where herbicides are currently used sparingly and farm wages are relatively low. However, this does not indicate a lack of importance for hand-weeding. Manual weeding is still the dominant weed control method in many parts of Asia, since management options for weed control are limited under diverse agro-ecological conditions (Kim, 2000).

All the recommended herbicides commonly used are known to be very safe not only for humans and cattle, but also for the environment when they are used properly. Sulfonylurea type herbicides have been intensively used in far-east Asia since 1990 because of their high efficacy against a broad spectrum of paddy weeds even at extremely low doses. However, intensive and repeated application of this type of herbicide has resulted in several negative effects, as follows:

- Evolving resistant weeds (Valverde et al. 2000)
- Residual effects on the following crops
- Disappearance of some susceptible weeds such as Brasenia schreberi and Sagittaria aginashi, which affects weed biodiversity (Itoh, 2000).

All these factors may well provide sufficient reason to attract public concern and anxiety regarding the negative effects of herbicides that might originate from intensive herbicide application in the environment. In this regard, an alternative to such a heavy dependence on herbicide is needed. Such an alternative might be found in the use of integrated weed management, which can reduce herbicide use in different cropping systems.

Organic Agriculture

During recent years, organic agriculture has hugely increased in many regions of the world, and this trend has not yet ceased. The total area of organically grown crops is estimated to be around over 8 million ha (reference) (Organic agriculture should be seen as a process depending highly on naturally-occurring biological processes. Therefore, the objective is to stimulate these processes to obtain maximum suppression of pest problems. Obviously, the use of chemical pesticides is contrary to the concept and the practice of organic agriculture, and human intervention in the process should therefore be carried out only very carefully.

Weed management in organic agriculture is not an easy task, particularly in areas where labour for hand-weeding is short or not affordable. However, the principle should be the same as in any conventional cropping system, i.e. weed competition needs to be prevented in order to obtain maximum crop yields. This necessarily implies weeding with non-chemical materials but, this has to be carried out exactly at the right time to eliminate weeds during the so-called 'critical period of weed competition. Organic systems also require the use of preventive methods before growing the crop and to establish a reasonable crop rotation. Stale seed bed preparation in order to kill the weeds mechanically or manually is a very good option to delay the start of weed competition. The use of cover crops and green manure, in addition to increasing soil fertility, may help to control some weed species. The most common methods used to prevent weed competition in organically grown crops are high seeding rates, narrow seed spacing/cross seeding, and companion cropping with small-seeded legumes. Rotational sequences for crops grown in tropical and sub-tropical areas should be developed to guarantee vigorous growth of the crops as well as fewer pest problems, including weeds.

Weed Control in Conservation Agriculture

The so-called Conservation Agriculture (CA) is gaining a lot of recognition among farmers all over the world. Grain crops as wheat, barley, maize, rice and soybean are grown in some parts of the world under this system, which consists of the judicious use of crop rotation with minimum or zero tillage, including the use of green manures and cover crops. This approach is beneficial to effectively protect and increase soil fertility. It is wrong to identify conservation agriculture as the practice of zero or minimum tillage. In fact these procedures are part of the system, but if they are implemented in areas of monocropping then this cannot be considered conservation agriculture. The use of herbicide-resistant crops (HRCs) combined with the application of broad-spectrum herbicides make the process of conservation agriculture easier, but it also runs the risk of bringing about new weed problems, either by a shift in the weed populations or the presence of species able to evolve resistance to the herbicides in use. Again it is not always necessary to use HRCs and herbicides, particularly when the system has been practised for ten years or more.

Although this development is real, several weeds still remain which seriously affect the production of a number of crops such as rice, with its incidence of red/weedy rice, and Echinochloa spp., parasitic Orobanche weeds found in sunflower, faba beans and solanaceous crops; parasitic Striga spp. in cereals in Africa south of the Sahara, and Imperata cylindrica in many areas in Africa.

The switch to zero tillage or direct seeding practices concerns many producers when it comes to controlling weeds. The loss of tillage as a method of weed control means that producers must adjust crop rotations, herbicide use, and other cultural practices to compensate. Perennial weeds may become a serious problem to overcome, and there is a need to implement additional cultural methods, such as the use of covers. Some annual weeds, such as wild oat and volunteer canola, also grow well under zero-till conditions.

Weeds exist in many different forms and with different life spans; there are annual, biennial, and perennial weeds. Weeds are not always bad, and in low density will result in no, or small yield losses. Heavy manifestations of weeds, or establishment of perennial weeds, can result in large yield losses or can even take land out of production until the weeds are controlled. Weeds can also reduce yield quality and can be toxic when ingested by animals or humans. Weeds can also cause environmental damage and loss of agricultural biodiversity, by competing for inputs. Integrated Weed Management aims at:

- Preventing weeds from spreading by:
  - Cleaning farm machinery and vehicles before transporting, to avoid risk of spreading weeds;
  - Cleaning the hair and feet of animals before moving to new areas;
  - Controlling the weeds in feed and bedding grounds;
  - Using only well stored and rotted manure (4-5 months), possibly improve decomposition;
  - Making sure that soil disturbances are immediately reseeded
  - When possible, practicing weed control on all aspects of the farm, including irrigation canals, drainage ditches, fence lines, stockyards, and farm roads.
- Improving knowledge of the identification and effects of different types of weeds.
- Monitoring and map the spread of weed populations and the resulting damage

- Making control decisions based on full knowledge of potential damage, cost of control methods, and the environmental impact of the control strategy.
- Using combinations of (preferably biological) weed control strategies to reduce the weed populations, these can include: winter cover crops, mulching, crop rotations, natural competition (e.g. ryegrass), livestock grazing, proper seedbed preparation, selecting locally adapted varieties, proper fertilizer application, stimulating bio-control by insects, mowing, hand weeding, and try to avoid, when possible herbicide application, tillage, and burning
- Evaluate and monitor the effectiveness and (environmental) effects of control strategies.

Mulch

A mulch is a layer of material applied to the surface of an area of soil. Its purpose is any or all of the following:-

- to conserve moisture
- to improve the fertility and health of the soil
- to reduce weed growth
- to enhance the visual appeal of the area

A mulch is usually but not exclusively organic in nature. It may be permanent (e.g. bark chips) or temporary (e.g. plastic sheeting). It may be applied to bare soil, or around existing plants. Mulches of manure or compost will be incorporated naturally into the soil by the activity of worms and other organisms. The process is used both in commercial crop production and in gardening, and when applied correctly can dramatically improve soil productivity.

Materials

Materials used as mulches vary and depend on a number of factors. Use takes into consideration availability, cost, appearance, the effect it has on the soil—including chemical reactions and pH, durability, combustibility, rate of decomposition, how clean it is—some can contain weed seeds or plant pathogens.

A variety of materials are used as mulch:

- Organic residues: grass clippings, leaves, hay, straw, kitchen scraps comfrey, shredded bark, whole bark nuggets, sawdust, shells, woodchips, shredded newspaper, cardboard, wool, animal manure, etc. Many of these materials also act as a direct composting system, such as the mulched clippings of a mulching lawn mower, or other organics applied as sheet composting.

- Compost: This should be fully composted material to avoid possible phytotoxicity problems, and the weed seed must have been eliminated, otherwise the mulch will actually produce weed cover.
- Rubber mulch: made from recycled tire rubber.
- Plastic mulch: crops grow through slits or holes in thin plastic sheeting. This method is predominant in large-scale vegetable growing, with millions of acres cultivated under plastic mulch worldwide each year (disposal of plastic mulch is cited as an environmental problem).
- Rock and gravel can also be used as a mulch. In cooler climates the heat retained by rocks may extend the growing season.

Organic Mulches

Organic mulches decay over time and are temporary. The way a particular organic mulch decomposes and reacts to wetting by rain and dew affects its usefulness.

Some mulches such as straw, peat and sawdust may negatively affect plant growth because of their wide carbon to nitrogen ratio, because bacteria and fungi remove nitrogen from the surrounding soil for growth. However, whether this effect has any practical impact on gardens is disputed by researchers and the experience of gardeners. Organic mulches can mat down, forming a barrier that blocks water and air flow between the soil and the atmosphere. Some organic mulches can wick water from the soil to the surface, which can dry out the soil.

Some organic mulches are coloured red, brown, black, and other colours. Isopropanolamine, specifically 1-Amino-2-propanol or DOW™ monoisopropanolamine, may be used as a pigment dispersant and colour fastener in these mulches.

Commonly available organic mulches include:

Leaves

- *Leaves* from deciduous trees, which drop their foliage in the autumn/fall. They tend to be dry and blow around in the wind, so are often chopped or shredded before application. As they decompose they adhere to each other but also allow water and moisture to seep down to the soil surface. Thick layers of entire leaves, especially of maples and oaks, can form a soggy mat in winter and spring which can impede the new growth lawn grass and other plants. Dry leaves are used as winter mulches to protect plants from freezing and thawing in areas with cold winters, they are normally removed during spring.

Grass Clippings

- *Grass clippings*, from mowed lawns are sometimes collected and used elsewhere as mulch. Grass clippings are dense and tend to mat down, so are mixed with tree leaves or rough compost to provide aeration and to facilitate their decomposition without smelly putrefaction. Rotting fresh grass clippings can damage plants; their rotting often produces a damaging buildup of trapped heat. Grass clippings are often dried thoroughly before application, which mediates against rapid decomposition and excessive heat generation. Fresh green grass clippings are relatively high in nitrate content, and when used as a mulch, much of the nitrate is returned to the soil, but the routine removal of grass clippings from the lawn results in nitrogen deficiency for the lawn.

Peat moss

- *Peat moss*, or sphagnum peat, is long lasting and packaged, making it convenient and popular as a mulch. When wetted and dried, it can form a dense crust that does not allow water to soak in. When dry it can also burn, producing a smoldering fire. It is sometimes mixed with pine needles to produce a mulch that is friable. It can also lower the pH of the soil surface, making it useful as a mulch under acid loving plants.

Wood Chips

- *Wood chips* are a byproduct of the pruning of trees by arborists, utilities and parks; they are used to dispose of bulky waste. Tree branches and large stems are rather coarse after chipping and tend to be used as a mulch at least three inches thick. The chips are used to conserve soil moisture, moderate soil temperature and suppress weed growth. The decay of freshly produced chips from recently living woody plants, consumes nitrate; this is often off set with a light application of a high-nitrate fertilizer. Wood chips are most often used under trees and shrubs. When used around soft stemmed plants, an unmulched zone is left around the plant stems to prevent stem rot or other possible diseases. They are often used to mulch trails, because they are readily produced with little additional cost outside of the normal disposal cost of tree maintenance. Wood chips come in various colours.

Woodchip Mulch

- *Woodchip mulch* is a byproduct of reprocessing used (untreated) timber (usually packaging pallets), to dispose of wood waste by creating woodchip mulch. The chips are used to conserve soil moisture, moderate soil temperature and suppress weed growth. Woodchip mulch is often used under trees, shrubs or large planting areas and can last much longer than arborist mulch. In addition, many consider woodchip mulch to be visually appealing, as it comes in various colours. Woodchips can also be reprocessed into playground woodchip to be used as an impact-attenuating playground surfacing.

Bark chips

- *Bark chips*, of various grades are produced from the outer corky bark layer of timber trees. Sizes vary from thin shredded strands to large coarse blocks. The finer types are very attractive but have a large exposed surface area that leads to quicker decay. Layers two or three inches deep are usually used, bark is relativity inert and its decay does not demand soil nitrates. Bark chips are also available in various colours.

Straw Mulch / field hay / Salt hay

- *Straw mulch* or *field hay* or *salt hay* are lightweight and normally sold in compressed bales. They have an unkempt look and are used in vegetable gardens and as a winter covering. They are biodegradable and neutral in pH. They have good moisture retention and weed controlling properties but also are more likely to be contaminated with weed seeds. Salt hay is less likely to have weed seeds than field hay. Straw mulch is also available in various colours.

Cardboard / Newspaper

- *Cardboard* or *newspaper* can be used as mulches. These are best used as a base layer upon which a heavier mulch such as compost is placed to prevent the lighter cardboard/newspaper layer from blowing away. By incorporating a layer of cardboard/ newspaper into a mulch, the quantity of heavier mulch can be reduced, whilst improving the weed suppressant and moisture retaining properties of the mulch. However, additional labour is expended when planting through a mulch containing a cardboard/newspaper layer, as holes must be cut for each plant. Sowing seed through mulches containing a cardboard/ newspaper layer is impractical. Application of newspaper mulch in windy weather can be facilitated by briefly pre-soaking the newspaper in water to increase its weight.

Coloured Mulch

Coloured mulch is mulch used for landscaping which is dyed in various colours, such as: red, black, gold, and shades of brown. Types of mulch which can be dyed include: wood chips, bark chips (barkdust) and pine straw.

Coloured mulch is made by dyeing the mulch in a water-based solution of colourant and chemical binder. When coloured mulch first entered the market, most formulas were suspected to contain toxic, heavy metals and other contaminates. Today, "current investigations indicate that mulch colourants pose no threat to people, pets or the environment. The dyes currently used by the mulch and soil industry are similar to those used in the cosmetic and other manufacturing industries (i.e., iron oxide)," as stated by the Mulch and Soil Council.

Coloured mulch can be applied anywhere non-coloured mulch is used (such as large bedded areas or around plants) and features many of the same gardening benefits as traditional mulch, such as improving soil productivity and retaining moisture.

As mulch decomposes, just as with non-coloured mulch, more mulch may need to be added to continue providing benefits to the soil and plants. However, if mulch is faded, spraying dye to previously spread mulch in order to restore colour is an option.

Application

Mulch is usually applied towards the beginning of the growing season, and is often reapplied as necessary. It serves initially to warm the soil by helping it retain heat which is lost during the night. This allows early seeding and transplanting of certain crops, and encourages faster growth. As the season progresses, mulch stabilizes the soil temperature and moisture, and prevents sunlight from germinating weed seeds.

In temperate climates, the effect of mulch is dependent upon the time of year at which it is applied as it tends to slow changes in soil temperature and moisture content. Mulch, when applied to the soil in late winter/early spring, will slow the warming of the soil by acting as an insulator, and will hold in moisture by preventing evaporation. Mulch, when applied at the time of peak soil temperatures in mid-summer, will maintain high soil temperatures further into the autumn (fall). The effect of mulch upon soil moisture content in mid-summer is complex however. Mulch prevents sunlight from reaching the soil surface, thus reducing evaporation. However, mulch can absorb much of the rainfall provided during light rainfall, which will later quickly evaporate when exposed to sunlight, thus preventing absorption into the soil, while heavy rainfall is able to saturate the mulch layer, and reach the soil below.

In order to maximise the benefits of mulch, while minimizing its negative influences, it is often applied in late spring/early summer when soil temperatures have risen sufficiently, but soil moisture content is still relatively high. Furthermore, at this point in the growing season, plants should be well enough established to be able to cope with the increase in the numbers of slugs and snails owing to the habitat provided for them by the mulch. However, permanent mulch is also widely used and valued for its simplicity, as popularized by author Ruth Stout, who said, "My way is simply to keep a thick mulch of any vegetable matter that rots on both sides of my vegetable and flower garden all year long. As it decays and enriches the soils, I add more."

Plastic mulch used in large-scale commercial production is laid down with a tractor-drawn or standalone layer of plastic mulch. This is usually part of a sophisticated mechanical process, where raised beds are formed, plastic is rolled out on top, and seedlings are transplanted through it. Drip irrigation is often required, with drip tape laid under the plastic, as plastic mulch is impermeable to water.

In home gardens and smaller farming operations, organic mulch is usually spread by hand around emerged plants. (On plots with existing mulch, the mulch is pulled away from the seedbed before planting, and restored after the seedlings have emerged.) For materials like straw and hay, a shredder may be used to chop up the material. Organic mulches are usually piled quite high, six inches (152 mm) or more, and settle over the season.

In some areas of the United States, such as central Pennsylvania and northern California, mulch is often referred to as "tanbark", even by manufacturers and distributors. In these areas, the word "mulch" is used specifically to refer to very fine tanbark or peat moss. Mulch made with wood can contain or feed termites, so care must be taken about not placing mulch too close to houses or building that can be damaged by those insects. Some mulch manufacturers recommend putting mulch several inches away from buildings.

Anaerobic (Sour) Mulch

Mulch normally smells like freshly cut wood, but sometimes develops a toxicity that causes it to smell like vinegar, ammonia, sulphur or silage. This happens when material with ample nitrogen content is not rotated often enough and it forms pockets of increased decomposition. When this occurs, the process may become anaerobic and produce these phytotoxic materials in small quantities. Once exposed to the air, the process quickly reverts to an aerobic process, but these toxic materials may be present for a period of time. If the mulch is placed around plants before the toxicity has had a chance to dissipate, then the plants could very likely be damaged or killed depending on their hardiness. Plants that are predominantly low to the ground or freshly planted are the most susceptible, and the phytotoxicity may prevent germination of some seeds.

If sour mulch is applied and there is plant kill, the best thing to do is to water the mulch heavily. Water dissipates the chemicals faster and refreshes the plants. Removing the offending mulch may have little effect, because by the time plant kill is noticed, most of the toxicity is already dissipated. While testing after plant kill will not likely turn up anything, a simple pH check may reveal high acidity, in the range of 3.8 to 5.6 instead of the normal range of 6.0 to 7.2. Finally, placing a bit of the offending mulch around another plant to check for plant kill will verify if the toxicity has departed. If the new plant is also killed, then sour mulch is probably not the problem.

Groundcovers (Living Mulches)

Groundcovers are plants which grow close to the ground, under the main crop, to slow the development of weeds and provide other benefits of mulch. They are usually fast-growing plants that continue growing with the main crops. By contrast, cover crops are incorporated into the soil or killed with herbicides. However, live mulches also may need to be mechanically or chemically killed eventually to prevent competition with the main crop.

Some groundcovers can perform additional roles in the garden such as nitrogen fixation in the case of clovers, dynamic accumulation of nutrients from the subsoil in the case of creeping comfrey (*Symphytum ibericum*), and even food production in the case of *Rubus Tricolor*.

On-site Mulch Production

Owing to the great bulk of mulch which is often required on a site, it is often impractical and expensive to source and import sufficient mulch materials. An alternative to importing mulch materials is to grow them on site in a "mulch garden" - an area of the site dedicated entirely to the production of mulch which is then transferred to the growing area. Mulch gardens should be sited as close as possible to the growing area so as to facilitate transfer of mulch materials.

Mulching (Composting) Over Unwanted Plants

Sufficient mulch over plants will destroy them, and may be more advantageous than using herbicide, cutting, mowing, pulling, raking, or tilling. The higher the temperature that this "mulch" is composted, the quicker the reduction of undesirable materials. "Undesirable materials" may include living seed, plant "trash", as well as pathogens such as from animal feces, urine (e.g. hantavirus), fleas, lice, ticks, etc.

In some ways this improves the soil by attracting and feeding earthworms, and adding humus. Earthworms "till" the soil, and their feces are among the best fertilizers and soil conditioners.

Urine may be toxic to plants if applied to growing areas undiluted.

Fertility (Soil)

Fertile soil has the following properties:

- It is rich in nutrients necessary for basic plant nutrition, including nitrogen,
- It contains sufficient minerals (trace elements) for plant nutrition, including boron, chlorine, cobalt, copper, iron, manganese, magnesium, molybdenum, sulphur, and zinc.
- It contains soil organic matter that improves soil structure and soil moisture retention.
- Soil pH is in the range 6.0 to 6.8 for most plants but some prefer acid or alkaline conditions.
- Good soil structure, creating well drained soil, but some soils are wetter (as for producing rice) or drier (as for producing plants susceptible to fungi or rot) such as agave.

- A range of microorganisms that support plant growth.
- It often contains large amounts of topsoil.

In lands used for agriculture and other human activities, fertile soil typically arises from the use of soil conservation practices. Basically the ability of a soil to supply plant nutrients.

Soil Fertilization

Bioavailable nitrogen is the element in soil that is most often lacking. Phosphorus and potassium are also needed in substantial amounts. For this reason these three elements are always identified on a commercial fertilizer analysis. For example a 10-10-15 fertilizer has 10 percent nitrogen, 10 percent (P_2O_5) available phosphorus and 15 percent (K_2O) water soluble potassium. Sulphur is the fourth element that may be identified in a commercial analysis—e.g. 21-0-0-24 which would contain 21% nitrogen and 24% sulphate.

Inorganic fertilizers are generally less expensive and have higher concentrations of nutrients than organic fertilizers. Some have criticized the use of inorganic fertilizers, claiming that the water-soluble nitrogen doesn't provide for the long-term needs of the plant and creates water pollution. Slow-release fertilizers may reduce leaching loss of nutrients and may make the nutrients that they provide available over a longer period of time.

Soil fertility is a complex process that involves the constant cycling of nutrients between organic and inorganic forms. As plant material and animal wastes decompose they release nutrients to the soil solution. Those nutrients may then undergo further transformations which may be aided or enabled by soil micro-organisms. Natural processes such as lightning strikes may fix atmospheric nitrogen by converting it to (NO_2). Denitrification may occur under anaerobic conditions (flooding) in the presence of denitrifying bacteria. The cations, primarily phosphate and potash, as well as many micronutrients are held in relatively strong bonds with the negatively charged portions of the soil in a process known as Cation Exchange Capacity

In 2008 the cost of phosphorus as fertilizer more than doubled, while the price of rock phosphate as base commodity rose eight-fold. Recently the term peak phosphorus has been coined, due to the limited occurrence of rock phosphate in the world.

Light and CO_2 Limitations

Photosynthesis is the process whereby plants use light energy to drive chemical reactions which convert CO_2 into sugars. As such, all plants require access to both light and carbon dioxide to produce energy, grow and reproduce.

While typically limited by nitrogen, phosphorus and potassium, low levels of carbon dioxide can also act as a limiting factor on plant growth. Peer-reviewed and published scientific studies have shown that increasing CO_2 is highly effective at promoting plant growth up to levels over 300 ppm. Further increases in CO_2 can, to a very small degree, continue to increase net photosynthetic output.

Since higher levels of CO_2 have only a minimal impact on photosynthetic output at present levels (presently around 400 ppm and increasing), we should not consider plant growth to be limited by carbon dioxide. Other biochemical limitations, such as soil organic content, nitrogen in the soil, phosphorus and potassium, are far more often in short supply. As such, neither commercial nor scientific communities look to air fertilization as an effective or economic method of increasing production in agriculture or natural ecosystems. Furthermore, since microbial decomposition occurs faster under warmer temperatures, higher levels of CO_2 (which is one of the causes of unusually fast climate change) should be expected to increase the rate at which nutrients are leached out of soils and may have a negative impact on soil fertility.

Soil Depletion

Soil depletion occurs when the components which contribute to fertility are removed and not replaced, and the conditions which support soil's fertility are not maintained. This leads to poor crop yields. In agriculture, depletion can be due to excessively intense cultivation and inadequate.

One of the most widespread occurrences of soil depletion as of 2008 is in tropical zones where nutrient content of soils is low. The combined effects of growing population densities, large-scale industrial logging, slash-and-burn agriculture and ranching, and other factors, have in some places depleted soils through rapid and almost total nutrient removal.

Topsoil depletion occurs when the nutrient-rich organic topsoil, which takes hundreds to thousands of years to build up under natural conditions, is eroded or depleted of its original organic material. Historically, many past civilizations' collapses can be attributed to the depletion of the topsoil. Since the beginning of agricultural production in the Great Plains of North America in the 1880s, about one-half of its topsoil has disappeared.

Depletion may occur through a variety of other effects, including overtillage (which damages soil structure), underuse of nutrient inputs which leads to mining of the soil nutrient bank, and salinization of soil.

Depleted soils can be improved over time through the use of conservation tillage practices, such as no-till and minimum-till, as well as by incorporating organic amendments like manure or compost into the soil. Inorganic fertilizer and mineral amendments such as lime are also used to help increase soil fertility.

Soil Testing and Fetilizer Use

Scope of Soil Science

Soil Science has six well defined and developed disciplines. Scope of soil Science is reflected through these disciplines.

Soil Science: The science dealing with soil as a natural resource on the surface of the earth, including Pedology (soil genesis, classification and mapping) and the physical, chemical and biological and fertility properties of soil and these properties in relation to their management for crop production.

1. Soil fertility: Nutrient supplying properties of soil
2. Soil chemistry: Chemical constituents, chemical properties and the chemical reactions
3. Soil physics: Involves the study of physical properties
4. Soil microbiology: deals with micro organisms, its population, classification, its role in transformations
5. Soil conservation: Dealing with protection of soil against physical loss by erosion or against chemical deterioration i.e. excessive loss of nutrients either natural or artificial means.
6. Pedology: Dealing with the genesis, survey and classification

Soil can be compared to various systems of human body

Digestive - matters decomposition

Respiratory - air circulation & exchange of gases

Circulatory - water movement with in the soil system

Excretory - leaching out of excess salts

Brain - soil clay

Colour - soil colour

Height - soil depth

Components of Soil (Volume basis)

Mineral matter – 45%

Organic matter – 5%

Soil water – 25%

Soil air – 25%

Definition, Objective and Philosophy of Soil Testing

Soil Testing: Soil testing is a rapid chemical analysis to access available nutrient status of the soil and includes interpretation, evaluation and fertilizer recommendation based on the result of chemical analysis and other considerations.

Objectives of Soil Testing:

1. Grouping of soil into classes relative to the nutrient level.
2. Predicating the probability of getting a profitable response to the fertilizers.
3. To provide the basis for fertilizer recommendation.

Philosophy of Soil Testing Programme:

1. To assess the adequacy of supply of each plant nutrients in the soil based on careful and thorough soil testing programme.
2. To test the use of plant nutrients based on following consideration.
    a. Application of plant nutrients only if the soil is inadequate or desired crop yield and Maintaince of soil supplies. Plant nutrient use is not recommended for crops on soils that able to supply adequate amount for the desired copy yields.
    b. The level of production desired or yield goal. Higher level of production require more nutrient at a given soil supplied than lower level of productions.
3. The predicting capability of soil testing programme is based on
    a. Calibration and correlation of soil analysis with crop yield research programme.
    b. Thorough and careful collection of soils samples to represent the area being considered.
    c. Accurate and reproducible laboratory measurement.

Weathering of Rocks and Minerals

Rocks and minerals are formed under a very high temperature and pressure, exposed to atmospheric conditions of low pressure and low temperature and they become unstable and weather. Soils are formed from rocks through the intermediate stage of formation of Regolith which is the resultant of weathering.

The Sequence of Processes in the Formation of Soils is: Weathering of rocks and minerals -> formation of regolith or parent material ->formation of true soil from regolith

Rock ->Weathering ->Regolith ->Soil forming factors and processes ->True soil (otherwise)

Two processes involved in the formation of soil are:

1. Formation of regolith by breaking down (weathering) of the bed rock.
2. The addition of organic matter through the decomposition of plant and animal tissues, and reorganization of these components by soil forming processes to form soil.

Weathering: A process of disintegration and decomposition of rocks and minerals which are brought about by physical agents and chemical processes, leading to the formation of Regolith (unconsolidated residues of the weathering rock on the earth's surface or above the solid rocks).

(OR)

The process by which the earth's crust or lithosphere is broken down by the activities of the atmosphere, with the aid of the hydrosphere and biosphere

(OR)

The process of transformation of solid rocks into parent material or Regolith

Parent material: It is the regolith or at least its upper portion. May be defined as the unconsolidated and more or less chemically weathered mineral materials from which soil are developed

Two basic processes of Weathering:

Physical (or) mechanical - disintegration

Chemical – decomposition

In addition, another process: Biological and all these processes are work hand in hand

Depending up on the agents taking part in weathering processes, it is classified into three types.

Different Agents of Weathering

Physical/ Mechanical (disintegration)	*Chemical (decomposition)*	*Biological (disint + decomp)*
1.Physical condition of rock	1.Hydration	1.Man & animals
2.Change in temperature	2.Hydrolysis	2. higher plants & their roots
3.Action of H2O	3.Solution	3.Micro organisms
-fragment & transport	4.Carbonation	
- action of freezing	5.Oxidation	
- alter. Wet & drying	6.Reduction	
- action of glaciers		
4.Action of wind		
5. Atmosp. electric pheno		

Physical Weathering of Rocks

The rocks are disintegrated and are broken down to comparatively smaller pieces, with out producing any new substances

Physical Condition of Rocks: The permeability of rocks is the most important single factor.

1. Coarse textured (porous) sand stone weather more readily than a fine textured (almost solid) basalt.
2. Unconsolidated volcanic ash weather quickly as compared to unconsolidated coarse deposits such as gravels

Action of Temperature: The variations in temperature exert great influence on the disintegration of rocks.

1. During day time, the rocks get heated up by the sun and expand. At night, the temperature falls and the rocks get cooled and contract.
2. This alternate expansion and contraction weakens the surface of the rock and crumbles it because the rocks do not conduct heat easily.
3. The minerals with in the rock also vary in their rate of expansion and contraction
4. The cubical expansion of quartz is twice as feldspar
5. Dark coloured rocks are subjected to fast changes in temperature as compared to light coloured rocks
6. The differential expansion of minerals in a rock surface generates stress between the heated surface and cooled unexpanded parts resulting in fragmentation of rocks.

7. This process causes the surface layer to peel off from the parent mass and the rock ultimately disintegrates. This process is called Exfoliation

Action of Water: Water acts as a disintegrating, transporting and depositing agent.

Fragmentation and Transport: Water beats over the surface of the rock when the rain occurs and starts flowing towards the ocean

1. Moving water has the great cutting and carrying force.
2. It forms gullies and ravines and carries with the suspended soil material of variable sizes.
3. Transporting power of water varies. It is estimated that the transporting power of stream varies as the sixth power of its velocity i.e the greater the speed of water, more is the transporting power and carrying capacity

Speed/Sec	***Carrying capacity***
15 cm	fine sand
30 cm	gravel
1.2 m	stones (1kg)
9.0 m	boulders (several tons)

The disintegration is greater near the source of river than its mouth

Action of Freezing: Frost is much more effective than heat in producing physical weathering

1. In cold regions, the water in the cracks and crevices freezes into ice and the volume increases to one tenth
2. As the freezing starts from the top there is no possibility of its upward expansion. Hence, the increase in volume creates enormous out ward pressure which breaks apart the rocks

Alternate Wetting and Drying: Some natural substances increase considerably in volume on wetting and shrink on drying.(e.g.) smectite, montmorilonite

1. During dry summer/ dry weather – these clays shrink considerably forming deep cracks or wide cracks.
2. On subsequent wetting, it swells.
3. This alternate swelling and shrinking/ wetting or drying of clay enriched rocks make them loose and eventually breaks

Action of glaciers:

1. In cold regions, when snow falls, it accumulates and changes into ice sheet.
2. These big glaciers start moving owing to the change in temperature and/or gradient.
3. On moving, these exert tremendous pressure over the rock on which they pass and carry the loose materials
4. These materials get deposited on reaching the warmer regions, where its movement stops with the melting of ice

Action of wind:

1. Wind has an erosive and transporting effect. Often when the wind is laden with fine material viz., fine sand, silt or clay particles, it has a serious abrasive effect and the sand laden winds itch the rocks and ultimately breaks down under its force
2. The dust storm may transport tons of material from one place to another. The shifting of soil causes serious wind erosion problem and may render cultivated land as degraded (e.g.) Rajasthan deserts

Atmospheric Electrical Phenomenon: It is an important factor causing break down during rainy season and lightning breaks up rocks and or widens cracks.

Chemical Weathering of Rocks

Decomposition of rocks and minerals by various chemical processes is called chemical weathering. It is the most important process for soil formation.

Chemical weathering takes place mainly at the surface of rocks and minerals with disappearance of certain minerals and the formation of secondary products (new materials). This is called chemical transformation.

Feldspar + water -> clay mineral + soluble cations and anions

Chemical weathering becomes more effective as the surface area of the rock increases.

Since the chemical reactions occur largely on the surface of the rocks, therefore the smaller the fragments, the greater the surface area per unit volume available for reaction.

The effectiveness of chemical weathering is closely related to the mineral composition of rocks. (e.g.) quartz responds far slowly to the chemical attack than olivine or pyroxene.

Average Mineralogical Composition (%)

Composition	*Granite*	*Basalt*	*Shale*	*S. Stone*	*L. Stone*
Feldspar	52.4	46.2	30.0	11.5	-
Quartz	31.3	-	2.3	66.8	-
Pyrox - amphi	-	44.5	-	-	-
FeO mineral	2.0	9.3	10.5	2.0	-
Clay mineral	14.3	-	25.0	6.6	24.0
Carbonates	-	-	5.7	11.1	76.0

Chemical Processes of weathering:

Hydration: Chemical combination of water molecules with a particular substance or mineral leading to a change in structure.

Soil forming minerals in rocks do not contain any water and they under go hydration when exposed to humid conditions. Up on hydration there is swelling and increase in volume of minerals. The minerals loose their luster and become soft.

It is one of the most common processes in nature and works with secondary minerals, such as aluminium oxide and iron oxide minerals and gypsum. (e.g.)

a) $2Fe_2O_3 + 3HOH \rightarrow 2Fe_2O_3 .3H_2O$

(Hematite) (Red) (Limonite) (Yellow)

b) $Al_2O_3 + 3HOH \rightarrow Al_2O_3 .3H_2O$

(Bauxite) (Hyd. aluminium Oxide)

c) $CaSO_4 + 2H_2O \rightarrow CaSO_4 .2H_2O$

(Anhydrite) (Gypsum)

d) $3(MgO.FeO.SiO_2) + 2H_2O \rightarrow 3MgO.2SiO_2.2H_2O + SiO_2 + 3H_2O$

(Olivine) (Serpentine)

Hydrolysis: Most important process in chemical weathering. It is due to the dissociation of H_2O into H^+ and OH^- ions which chemically combine with minerals and bring about changes, such as exchange, decomposition of crystalline structure and formation of new compounds. Water acts as a weak acid on silicate minerals.

$KAlSi_3O_8 + H_2O \rightarrow HAlSi_3O_8 + KOH$

(Orthoclase) (Acid silt clay)

$HAlSi_3O_8 + 8\ HOH \rightarrow Al_2O_3 .3H_2O + 6\ H_2SiO_3$

(Recombination) (Hyd. Alum. oxide) (Silicic acid)

This reaction is important because of two reasons.

1. clay, bases and Silicic acid - the substances formed in these reactions - are available to plants

2. water often containing CO_2 (absorbed from atmosphere), reacts with the minerals directly to produce insoluble clay minerals, positively charged metal ions (Ca^{++}, Mg^{++}, Na^+, K^+) and negatively charged ions (OH^-, HCO_3^-) and some soluble silica – all these ions are made available for plant growth.
3. Solution: Some substances present in the rocks are directly soluble in water. The soluble substances are removed by the continuous action of water and the rock no longer remains solid and form holes, rills or rough surface and ultimately falls into pieces or decomposes. The action is considerably increased when the water is acidified by the dissolution of organic and inorganic acids. (e.g) halites, NaCl $NaCl + H_2O \rightarrow Na^+, Cl^-, H_2O$ (dissolved ions with water)
4. Carbonation: Carbon dioxide when dissolved in water it forms carbonic acid. $2H_2O + CO_2 \rightarrow H_2CO_3$

This carbonic acid attacks many rocks and minerals and brings them into solution. The carbonated water has an etching effect up on some rocks, especially lime stone. The removal of cement that holds sand particles together leads to their disintegration.

$CaCO_3 + H_2CO_3 \rightarrow Ca(HCO_3)_2$

(Calcite) slightly soluble (Ca bi carbonate) readily soluble

5. Oxidation: The process of addition and combination of oxygen to minerals. The absorption is usually from O_2 dissolved in soil water and that present in atmosphere. The oxidation is more active in the presence of moisture and results in hydrated oxides.(e.g) minerals containing Fe and Mg.

$4FeO$ (Ferrous oxide) + O_2 $\rightarrow$ $2Fe_2O_3$ (Ferric oxide)

$4Fe_3O_4$ (Magnetite) + O_2 $\rightarrow$ $6Fe_2O_3$ (Hematite)

$2Fe_2O_3$ (Hematite) + $3H_2O$ $\rightarrow$ $2Fe_2O_3 .3H_2O$(Limonite)

6. Reduction: The process of removal of oxygen and is the reverse of oxidation and is equally important in changing soil colour to grey, blue or green as ferric iron is converted to ferrous iron compounds. Under the conditions of excess water or water logged condition (less or no oxygen), reduction takes place.

$2Fe_2O_3$ (Hematite) - O_2 $\rightarrow$ $4FeO$(Ferrous oxide) - reduced form

In conclusion, during chemical weathering igneous and metamorphic rocks can be regarded as involving destruction of primary minerals and the production of secondary minerals.

In sedimentary rocks, which is made up of primary and secondary minerals, weathering acts initially to destroy any relatively weak bonding agents (FeO) and the particles are freed and can be individually subjected to weathering.

Biological Weathering of Rocks

Unlike physical and chemical weathering, the biological or living agents are responsible for both decomposition and disintegration of rocks and minerals. The biological life is mainly controlled largely by the prevailing environment.

Man and Animals:

1. The action of man in disintegration of rocks is well known as he cuts rocks to build dams, channels and construct roads and buildings. All these activities result in increasing the surface area of the rocks for attack of chemical agents and accelerate the process of rock decomposition.
2. A large number of animals, birds, insects and worms, by their activities they make holes in them and thus aids for weathering.
3. In tropical and sub tropical regions, ants and termites build galleries and passages and carry materials from lower to upper surface and excrete acids. The oxygen and water with many dissolved substances, reach every part of the rock through the cracks, holes and galleries, and thus brings about speedy disintegration.
4. Rabbits, by burrowing in to the ground, destroy soft rocks. Moles, ants and bodies of the dead animals, provides substances which react with minerals and aid in decaying process
5. The earthworms pass the soil through the alimentary canal and thus bring about physical and chemical changes in soil material.

Higher Plants and Roots: The roots of trees and other plants penetrates into the joints and crevices of the rocks. As they grew, they exert a great disruptive force and the hard rock may break apart. (e.g.) pipal tree growing on walls/ rocks

The grass root form a sponge like mass prevents erosion and conserve moisture and thus allowing moisture and air to enter in to the rock for further action.

Some roots penetrate deep into the soil and may open some sort of drainage channel. The roots running in crevices in lime stone and marble produces acids. These acids have a solvent action on carbonates.

The dead roots and plant residues decompose and produce carbon dioxide which is of great importance in weathering.

Micro- organisms: In early stages of mineral decomposition and soil formation, the lower forms of plants and animals like, mosses, bacteria and fungi and actinomycetes play an important role. They extract nutrients from the rock and N from air and live with a small quantity of water. In due course of time, the soil develops under the cluster of these micro-organisms.

This organism closely associated with the decay of plant and animal remains and thus liberates nutrients for the use of next generation plants and also produces CO2 and organic compounds which aid in mineral decomposition.

Weathering of Minerals

There are many factors which influence the weathering of minerals.

1. Climatic conditions
2. Physical characteristics
3. Chemical and structural characteristics

Climatic Conditions: The climatic condition, more than any other factor tends to control the kind and rate of weathering. Under conditions of low rainfall, there is a dominance of physical weathering which reduces the size and increases the surface area with little change in volume.

The increase in moisture content encourages chemical as well as mechanical changes and new minerals and soluble products are formed.

The weathering rates are generally fastest in humid tropical regions as there is sufficient moisture and warmth to encourage chemical decomposition.

The easily weather able minerals disappear on account of intense chemical weathering and more resistant products (hydrous oxides of Fe and Al) tend to accumulate

Climate controls the dominant type of vegetation which in turn controls the bio chemical reactions in soils and mineral weathering.

Physical Characteristics:

i) Differential composition

ii) Particle size

iii) Hardness and degree of cementation

Chemical and Structural Characteristics: Chemical: For minerals of given particle size, chemical and crystalline characteristics determine the ease of decomposition. (e.g.) gypsum – sparingly soluble in water, is dissolved and removed in solution form under high rainfall.

Ferro magnesium minerals are more susceptible to chemical weathering than feldspar and quartz

Tightness of packing of ions in crystals: Less tightly packed minerals like olivine and biotitic are easily weathered as compared to tightly packed zircon and muscovite (resistant)

Chemical Weathering of Silicates

The most important silicates are quartz, feldspar and certain Ferro magnesium minerals

Weathering products of common silicate minerals

Minerals	*Composition*	*Decomposition products*	
		Minerals	*Others*
Olivine	(Fe, Mg)2SiO2	serpentine, limonitehaematite, quartz	Some Si in solution, carbonates of Fe and Mg
Pyroxenes	Fe, Mg	Clay, calcite, limonite	Some Si in solution, carbonates of Ca and Mg
Amphibole	Ca- silicates	Haematite, quartz	-do -
Biotite	Al		-do -
Plagioclase	Calcic	Clay, quartz, calcite	Some Si in solution, Na and Ca carbonates
	Sodic		-do-
Orthoclase	Potassic	Clay, quartz	Some Si in solution, potassium carbonate
Quartz		Quartz grains	Some Si in solution

Soil Forming Factors

The soil formation is the process of two consecutive stages.

1. The weathering of rock (R) into Regolith
2. The formation of true soil from Regolith

The evolution of true soil from regolith takes place by the combined action of soil forming factors and processes.

1. The first step is accomplished by weathering (disintegration & decomposition)
2. The second step is associated with the action of Soil Forming Factors

Weathering Factors

Dokuchaiev (1889) established that the soils develop as a result of the action of soil forming factors

S = f (P, Cl, O)

Further, Jenny (1941) formulated the following equation

S = f (Cl, O, R, P, T, ...)

Where,

Cl – environmental climate

o – Organisms and vegetation (biosphere)

r – Relief or topography

p – Parent material

t- Time

... - additional unspecified factors

The five soil forming factors, acting simultaneously at any point on the surface of the earth, to produce soil

Two groups – Passive i) Parent material, ii) Relief, iii) Time

Active IV) Climate, v) Vegetation & organism

Passive Soil Forming Factors

The passive soil forming factors are those which represent the source of soil forming mass and conditions affecting it. These provide a base on which the active soil forming factors work or act for the development of soil.

Parent Material: It is that mass (consolidated material) from which the soil has formed.

Two Groups of Parent Material

i) Sedentary: Formed in original place. It is the residual parent material. The parent material differ as widely as the rocks

ii) Transported: The parent material transported from their place of origin. They are named according to the main force responsible for the transport and redeposition.

    a) By gravity - Colluvial

    b) By water - Alluvial, Marine, Locustrine

    c) By ice - Glacial

    d) By wind – Eolian

Colluvium: It is the poorly sorted materials near the base of strong slopes transported by the action of gravity.

Alluvium: The material transported and deposited by water is, found along major stream courses at the bottom of slopes of mountains and along small streams flowing out of drainage basins.

Locustrine: Consists of materials that have settled out of the quiet water of lakes.

Moraine: Consists of all the materials picked up, mixed, disintegrated, transported and deposited through the action of glacial ice or of water resulting primarily from melting of glaciers.

Loess or Aeolian: These are the wind blown materials.

When the texture is silty - loess; when it is sand - Eolian.

The soils developed on such transported parent materials bear the name of the parent material; viz. Alluvial soils from alluvium, Colluvial soils from Colluvium etc. In the initial stages, however, the soil properties are mainly determined by the kind of parent material.

Endodynamomorphic soils: With advanced development and excessive leaching, the influence of parent material on soil characteristics gradually diminishes. There are soils wherein the composition of parent material subdues the effects of climate and vegetation. These soils are temporary and persist only until the chemical decomposition becomes active under the influence of climate and vegetation.

Ectodynamomorphic soils: Development of normal profile under the influence of climate and vegetation.

Soil properties as influenced by parent material: Different parent materials affect profile development and produce different soils, especially in the initial stages.

1. Acid igneous rocks (like granite, rhyolite) produce light-textured soils (Alfisols).
2. Basic igneous rocks (basalt), alluvium or Colluvium derived from limestone or basalt, produce fine-textured cracking-clay soils (Vertisols).
3. Basic alluvium or Aeolian materials produce fine to coarse-textured soils (Entisols or Inceptisols).
4. The nature of the elements released during the decaying of rocks has a specific role in soil formation. (e.g.) Si and Al form the skeleton for the production of secondary clay minerals.

5. Iron and manganese are important for imparting red colour to soils and for oxidation and reduction phenomena.
6. Sodium and potassium are important dispersing agents for day and humus colloids.
7. Calcium and magnesium have a flocculating effect and result in favourable and stable soil structure for plant growth.

Relief or Topography: The relief and topography sometimes are used as synonymous terms. They denote the configuration of the land surface. The topography refers to the differences in elevation of the land surface on a broad scale.

The prominent types of topography designations, as given in FAO Guidelines (1990) are:

Sr. No	***Land surface***	***with slopes of***
1	Flat to Almost flat	0 – 2 %
2	Gently undulating	2 - 5 %
3	Undulating	5 – 10 %
4	Rolling	10 – 15 %
5	Hilly	15 –3 0 %
6	Steeply dissect	>30% with moderate range of elevation(<300 m)
7	Mountainous	> 30% with great range of elevation(>300 m)

Soil formation on flat to almost flat position: On level topographic positions, almost the entire water received through rain percolates through the soil. Under such conditions, the soils formed may be considered as representative of the regional climate. They have normal solum with distinct horizons. But vast and monotonous level land with little gradient often has impaired drainage conditions.

Soil Formation on Undulating Topography: The soils on steep slopes are generally shallow, stony and have weakly- developed profiles with less distinct horizonation. It is due to accelerated erosion, which removes surface material before it has the time to develop. Reduced percolation of water through soil is because of surface runoff, and lack of water for the growth of plants, which are responsible for checking of erosion and promote soil formation.

Soil Formation in Depression: The depression areas in semi-arid and sub humid regions reflect more moist conditions than actually observed on level topographic positions due to the additional water received as runoff. Such conditions (as in the Tarai region of the Uttar

Pradesh) favour more vegetative growth and slower rate of decay of organic remains. This results in the formation of comparatively dark-coloured soils rich in organic matter (Mollisols).

Soil Formation and Exposure/ Aspect: Topography affects soil formation by affecting temperature and vegetative growth through slope exposures (aspect}. The southern exposures (facing the sun) are warmer and subject to marked fluctuations in temperature and moisture. The northern exposures, on the other hand are cooler and more humid. The eastern and western exposures occupy intermediate position in this respect.

Time: Soil formation is a very slow process requiring thousands of years to develop a mature pedon. The period taken by a given soil from the stage of weathered rock (i.e. regolith) up to the stage of maturity is considered as time. The matured soils mean the soils with fully developed horizons (A, B, C). It takes hundreds of years to develop an inch of soil. The time that nature devotes to the formation of soils is termed as Pedological Time.

It has been observed that rocks and minerals disintegrate and/ or decompose at different rates; the coarse particles of limestone are more resistant to disintegration than those of sandstone. However, in general, limestone decomposes more readily than sandstone (by chemical weathering).

Weathering stages in soil formation

Sr. No	***Stages***	***Characteristic***
1	Initial	Un weathered parent material
2	Juvenile	Weathering started but much of the original material still un weathered
3	Virile	Easily weather able minerals fairly decomposed; clay content increased, slowly weather able minerals still appreciable
4	Senile	Decomposition reaches at a final stage; only most resistant minerals survive
5	Final	Soil development completed under prevailing environments

1. The soil properties also change with time, for instance nitrogen and organic matter contents increase with time provided the soil temperature is not high.
2. CaCO3 content may decrease or even lost with time provided the climatic conditions are not arid
3. In humid regions, the H+ concentration increases with time because of chemical weathering.

Active Soil Forming Factors

The active soil forming factors are those which supply energy that acts on the mass for the purpose of soil formation. These factors are climate and vegetation (biosphere).

1. Climate: Climate is the most significant factor controlling the type and rate of soil formation. The dominant climates recognized are:
    1. Arid climate: The precipitation here is far less than the water-need. Hence the soils remain dry for most of the time in a year.
    2. Humid climate: The precipitation here is much more than the water need. The excess water results in leaching of salt and bases followed by translocation of clay colloids.
    3. Oceanic climate: Moderate seasonal variation of rainfall and temperature.
    4. Mediterranean climate: The moderate precipitation. Winters and summers are dry and hot.
    5. Continental climate: Warm summers and extremely cool or cold winters.
    6. Temperate climate: Cold humid conditions with warm summers.
    7. Tropical and subtropical climate: Warm to hot humid with isothermal conditions in the tropical zone.

Climate Affects the Soil Formation Directly and Indirectly.

Directly, climate affects the soil formation by supplying water and heat to react with parent material.

Indirectly, it determines the fauna and flora activities which furnish a source of energy in the form of organic matter. This energy acts on the rocks and minerals in the form of acids, and salts are released. The indirect effects of climate on soil formation are most clearly seen in the relationship of soils to vegetation.

Precipitation and temperature are the two major climatic elements which contribute most to soil formation.

Precipitation: Precipitation is the most important among the climatic factors. As it percolates and moves from one part of the parent material to another. It carries with it substances in solution as well as in suspension. The substances so carried are re deposited in another part or completely removed from the material through percolation when the soil moisture at the surface evaporates causing an upward movement of water. The soluble substances move with it and are translocated to the upper layer. Thus rainfall brings about a redistribution of substances both soluble as well as in suspension in soil body.

Temperature:

1. Temperature is another climatic agent influencing the process of soil formation.
2. High temperature hinders the process of leaching and causes an upward movement of soluble salts.
3. High temperature favours rapid decomposition of organic matter and increase microbial activities in soil while low temperatures induce leaching by reducing evaporation and there by favour the accumulation of organic matter by slowing down the process of decomposition. Temperature thus controls the rate of chemical and biological reactions taking place in the parent material.

Jenney (1941) computed that in the tropical regions the rate of weathering proceeds three times faster than in temperate regions and nine times faster than in arctic.

Organism & Vegetation

Organism:

1. The active components of soil ecosystem are plants, animals, microorganisms and man.
2. The role of microorganisms in soil formation is related to the humification and mineralization of vegetation
3. The action of animals especially burrowing animals to dig and mix-up the soil mass and thus disturb the parent material
4. Man influences the soil formation through his manipulation of natural vegetation, agricultural practices etc.
5. Compaction by traffic of man and animals decrease the rate of water infiltration into the soil and thereby increase the rate of runoff and erosion.

Vegetation:

1. The roots of the plants penetrate into the parent material and act both mechanically and chemically.
2. They facilitate percolation and drainage and bring about greater dissolution of minerals through the action of CO2 and acidic substances secreted by them.

3. The decomposition and humification of the materials further adds to the solubilization of minerals
4. Forests – reduces temperature, increases humidity, reduce evaporation and increases precipitation.
5. Grasses reduce runoff and result greater penetration of water in to the parent material.

Soil Forming Processes

The pedogenic processes, although slow in terms of human life, yet work faster than the geological processes in changing lifeless parent material into true soil full of life.

- The pedogenic processes are extremely complex and dynamic involving many chemical and biological reactions, and usually operate simultaneously in a given area.
- One process may counteract another, or two different processes may work simultaneously to achieve the same result.
- Different processes or combination of processes operate under varying natural environment.

The collective interaction of various soil forming factors under different environmental conditions set a course to certain recognized soil forming processes.

The basic process involved in soil formation (Simonson, 1959) includes the following.

- Gains or Additions of water, mostly as rainfall, organic and mineral matter to the soil.
- Losses of the above materials from the soil.
- Transformation of mineral and organic substances within the soil.
- Translocation or the movement of soil materials from one point to another within the soil. It is usually divided into
  1. movement of solution (leaching) and
  2. movement in suspension (eluviation) of clay, organic matter and hydrous oxides

7

Fundamental Soil Forming Processes

Humification: Humification is the process of transformation of raw organic matter into humus. It is extremely a complex process involving various organisms.

First, simple compounds such as sugars and starches are attacked followed by proteins and cellulose and finally very resistant compounds, such as tannins, are decomposed and the dark coloured substance, known as humus, is formed.

Eluviation: It is the mobilization and translocation of certain constituent's viz. Clay, Fe2O3, Al2O3, SiO2, humus, CaCO3, other salts etc. from one point of soil body to another. Eluviation means washing out. It is the process of removal of constituents in suspension or solution by the percolating water from the upper to lower layers. The eluviation encompasses mobilization and translocation of mobile constituents resulting in textural differences. The horizon formed by the process of eluviation is termed as eluvial horizon (A2 or E horizon).

Translocation depends upon relative mobility of elements and depth of percolation.

Illuviation: The process of deposition of soil materials (removed from the eluvial horizon) in the lower layer (or horizon of gains having the property of stabilizing translocated clay materials) is termed as Illuviation. The horizons formed by this process are termed as illuvial horizons (B-horizons, especially Bt) The process leads to textural contrast between E and Bt horizons, and higher fine: total clay ratio in the Bt horizon.

Horizonation: It is the process of differentiation of soil in different horizons along the depth of the soil body. The differentiation is due to the fundamental processes, humification, eluviation and Illuviation.

Specific Soil Forming Processes

The basic pedologic processes provide a framework for later operation of more specific processes

Calcification

It is the process of precipitation and accumulation of calcium carbonate ($CaCO_3$) in some part of the profile. The accumulation of $CaCO_3$ may result in the development of a calcic horizon. Calcium is readily soluble in acid soil water and/or when CO_2 concentration is high in root zone as:

$$CO_2 + H_2O \longrightarrow H_2CO_3$$

$$H_2CO_3 + Ca \longrightarrow Ca(HCO_3)_2 \text{ (soluble)}$$

$$Ca(HCO_3)_2 \xrightarrow[CO_2]{Temp.} CaCO_3 + H_2O + CO_2 \text{ (precipitates)}$$

The process of precipitation after mobilization under these conditions is called calcification and the resulting illuviated horizon of carbonates is designated as Bk horizon (Bca).

Decalcification: It is the reverse of calcification that is the process of removal of $CaCO_3$ or calcium ions from the soil by leaching

$$\underset{\text{(insoluble)}}{CaCO_3} + CO_2 + H_2O \xrightarrow[CO_2]{Temp.} Ca(HCO_3)_2 \text{ (soluble)}$$

Podzolization

It is a process of soil formation resulting in the formation of Podzols and Podzolic soils. In many respects, podzolization is the negative of calcification. The calcification process tends to concentrate calcium in the lower part of the B horizon, whereas podzolization leaches the entire solum of calcium carbonates.

Apart from calcium, the other bases are also removed and the whole soil becomes distinctly acidic. In fact, the process is essentially one of acid leaching. The process operates under favourable combination of the following environments.

i) Climate: A cold and humid climate is most favourable for podzolization

ii) Parent material: Siliceous (Sandy) material, having poor reserves of weather able minerals, favour the operation of podzolization as it helps in easy percolation of water.

iii) Vegetation: Acid producing vegetation such as coniferous pines is essential

iv) Leaching and Translocation of Sesquioxide: In the process of decomposition of organic matter various organic acids are produced. The organic acids thus formed act with Sesquioxide and the remaining clay minerals, forming organic- Sesquioxide and organic clay complexes, which are soluble and move with the percolating water to the lower horizons (Bh, Bs). Aluminium ions in a water solution hydrolyze and make the soil solution very acidic.

2Al +6H2O à2 Al(OH)3 + 6H+

As iron and aluminium move about, the A horizon gives a bleached grey or ashy appearance. The Russians used the term Podzols (pod means under, the Zola means ash like i.e. ash-like horizon appearing beneath the surface horizon) for such soils.

To conclude, the Podzolization is a soil forming process which prevails in a cold and humid climate where coniferous and acid forming vegetations dominate. The humus and Sesquioxide become mobile and leached out from the upper horizons and deposited in the lower horizon.

Laterization: The term laterite is derived from the word later meaning brick or tile and was originally applied to a group of high clay Indian soils found in Malabar hills of Kerala, Tamil Nadu, Karnataka and Maharashtra.

It refers specifically to a particular cemented horizon in certain soils which when dried, become very hard, like a brick. Such soils (in tropics) when massively impregnated with Sesquioxide (iron and aluminium oxides) to extent of 70 to 80 per cent of the total mass, are called laterite or latosols (Oxisols). The soil forming process is called Laterization or Latozation.

Laterization is the process that removes silica, instead of sesquioxides from the upper layers and thereby leaving sesquioxides to concentrate in the solum. The process operates under the following conditions.

i) Climate: Unlike podzolization, the process of laterization operates most favourable in warm and humid (tropical) climate with 2000 to 2500 mm rainfall and continuous high temperature (25°C) throughout the year.

ii) Natural vegetation: The rain forests of tropical areas are favourable for the process.

iii) Parent Material: Basic parent materials, having sufficient iron bearing ferromagnesian minerals (Pyroxene, amphiboles, biotite and chlorite), which on weathering release iron, are congenial for the development of laterites.

Gleization: The term *glei* is of Russian origin means blue, grey or green clay. The Gleization is a process of soil formation resulting in the development of a glei (or gley horizon) in the lower part of the soil profile above the parent material due to poor drainage condition (lack of oxygen) and where waterlogged conditions prevail. Such soils are called hydro orphic soils.

The process is not particularly dependent on climate (high rainfall as in humid regions) but often on drainage conditions.

The poor drainage conditions result from:

1. Lower topographic position, such as depression land, where water stands continuously at or close to the surface.
2. Impervious soil parent material, and.
3. Lack of aeration.

Under such conditions, iron compounds are reduced to soluble ferrous forms. The reduction of iron is primarily biological and requires both organic matter and microorganisms capable of respiring anaerobically. The solubility of Ca, Mg, Fe, and Mn is increased and most of the iron exists as Fe^{++} organo complexes in solution or as mixed precipitate of ferric and ferrous hydroxides.

This is responsible for the production of typical bluish to grayish horizon with mottling of yellow and or reddish brown colours.

Salinization: It is the process of accumulation of salts, such as sulphates and chlorides of calcium, magnesium, sodium and potassium, in soils in the form of a salty (salic) horizon. It is quite common in arid and semi arid regions. It may also take place through capillary rise of saline ground water and by inundation with seawater in marine and coastal soils. Salt accumulation may also result from irrigation or seepage in areas of impeded drainage.

Desalinization: It is the removal by leaching of excess soluble salts from horizons or soil profile (that contained enough soluble salts to impair the plant growth) by ponding water and improving the drainage conditions by installing artificial drainage network.

Solonization or Alkalization: The process involves the accumulation of sodium ions on the exchange complex of the clay, resulting in the formation of sodic soils (Solonetz). All cations in solution are engaged in a reversible reaction with the exchange sites on the clay and organic matter particles.

The reaction can be represented as:

Ca.Mg.2NaX à Ca++ +Mg++ +2Na+ + x-6 +3CO3 2- à Na2CO3 + MgCO3 +CaCO3

(Where X represents clay or organic matter exchange sites)

Solodization or dealkalization: The process refers to the removal of Na+ from the exchange sites. This process involves dispersion of clay. Dispersion occurs when Na+ ions become hydrated. Much of the dispersion can be eliminated if Ca++ and or Mg++ ions are concentrated in the water, which is used to leach the soonest. These Ca and Mg ion can replace the Na on exchange complex, and the salts of sodium are leached out as:

2NaX + CaSO4 à Na2SO4 + CaX

(leachable)

10. Pedoturbation: Another process that may be operative in soils is pedoturbation. It is the process of mixing of the soil. Mixing to a certain extent takes place in all soils. The most common types of pedoturbation are:

1. Faunal pedoturbation: It is the mixing of soil by animals such as ants, earthworms, moles, rodents, and man himself
2. Floral pedoturbation : It is the mixing of soil by plants as in tree tipping that forms pits and mounds
3. Argillic pedoturbation: It is the mixing of materials in the solum by the churning process caused by swell shrink clays as observed in deep Black Cotton Soils.

Development of Soil Profile

The development of soil profile is a constructive process where in disintegrated material resulted from weathering of rocks and minerals gets converted into a soil body.

Definition of soil profile: The vertical section of the soil showing the various layers from the surface to the unaffected parent material is known as a soil profile.

The various layers are known as horizons. A soil profile contains three main horizons.

They are named as horizon A, horizon B and horizon C.

- The surface soil or that layer of soil at the top which is liable to leaching and from which some soil constituents have been removed is known as horizon A or the horizon of eluviation.
- The intermediate layer in which the materials leached from horizon A have been re-deposited is known as horizon B or the horizon of illuviation.
- The parent material from which the soil is formed is known as horizon C.

A Study of soil profile is important as it is historic record of all the soil forming processes and it forms the basis for the study in pedagogical investigations.

Soil profile is the key for the soil classification and also forms the basis for the practical utility of soils.

A hypothetical mineral soil profile will include O, A, B, C and R master horizons and all the possible sub-horizons.

Master Horizons and Sub Horizons

O horizon: It is called as organic horizon. It is formed in the upper part of the mineral soil, dominated by fresh or partly decomposed organic materials.

- This horizon contains more than 30% organic matter if mineral fraction has more than 50 % clay (or) more than 20 % organic matter if mineral fraction has less clay.
- The organic horizons are commonly seen in forest areas and generally absent in grassland, cultivated soils.
  - O1 - Organic horizon in which the original forms of the plant and animal residues can be recognized through naked eye.
  - O2 - Organic horizon in which the original plant or animal matter can not be recognized through naked eye.
- A horizon - Horizon of organic matter accumulation adjacent to surface and that has lost clay, iron and Aluminium.
  - A1 - Top most mineral horizon formed adjacent to the surface. There will be accumulation of humified organic matter associated with mineral fraction and darker in Colour than that of lower horizons due to organic matter.
  - A2 - Horizon of maximum eluviation of clay, iron and aluminium oxides and organic matter. Loss of these constituents generally results in accumulation of quartz

and other sand and silt size resistant minerals. Generally lighter in Colour than horizons above and below.

  - A3 - A transitional layer between A and B horizons with more dominated properties of A1 or A2 above than the underlying B horizon. This horizon is sometimes absent. Solum.

- B horizon - Horizon in which the dominant features are accumulation of clay, iron, aluminium or humus alone or in combination. Coating of sesquioxides will impart darker, stronger of red Colour than overlying or underlying horizons.
  - B1 - A transitional layer between A and B. More like A than B.
  - B2 - Zone of maximum accumulation of clay, iron and aluminium oxide that may have moved down from upper horizons or may have formed in situ. The organic matter content is generally higher and Colour darker than that of A2 horizon above.
  - B3 - Transitional horizon between B and C and with properties more similar to that of overlying B2 than underlying C.
- C horizon - It is the horizon below the solum (A + B), relatively less affected by soil forming processes. It is outside the zone of major biological activity. It may contain accumulation of carbonates or sulphates, calcium and magnesium
- R - Underlying consolidated bed rock and it may or may not be like the parent rock from which the solum is formed.

Besides, lower case letters are used to indicate the special features of master horizons. This case letters follow the subdivisions of master horizons. E.g. Ap - ploughed layer, B2t - illuvial clay

When two or more genetically unrelated (contrasting) materials are present in a profile as in the case of alluvial or colluvial soils then the phenomenon is known as lithological discontinuity. This is indicated by the use of Roman letters as prefixes to the master horizons. E.g. Ap, B2, II B22, IIIC.

Special Features: Soil Individual or Polypedon: The Soil Survey Staff (1960) defined the soil individual or polypedon (Pedon, Ground) as a natural unit of soil that differs from its adjoining unit on the landscape in one or more properties.

The term pedon has been proposed for small basic soil entities that are part of the continuum mantling the land.

A pedon is the smallest volume that can be called "a soil". The set of pedon must fit within the range of one series and occur in a contiguous group to form a polypedon.

A polypedon is therefore, defined as a contiguous similar pedons bounded on all sides by "not-soil or by pedons of unlike characters. It is a real physical soils body which has a minimum area of more than 1 sq. km and an unspecified maximum area.

Physical Properties of Soil

Physical properties (mechanical behaviour) of a soil greatly influence its use and behaviour towards plant growth. The plant support, root penetration, drainage, aeration, retention of moisture, and plant nutrients are linked with the physical condition of the soil. Physical properties also influence the chemical and biological behaviour of soil. The physical properties of a soil depend on the amount, size, shape, arrangement and mineral composition of its particles. These properties also depend on organic matter content and pore spaces

Important physical properties of soils:

1. Soil texture
2. Soil structure
3. Surface area
4. Soil density,
5. Soil porosity
6. Soil colour
7. Soil consistence.

Soil Texture

Definition of Soil Texture: Soil texture refers to the relative proportion of particles or it is the relative percentage by weight of the three soil separates viz., sand, silt and clay or simply refers to the size of soil particles.

The proportion of each size group in a given soil (the texture) can not be easily altered and it is considered as a basic property of a soil.

The soil separates are defined in terms of diameter in millimeters of the particles. Soil particles less than 2 mm in diameter are excluded from soil textural determinations.

Stones and gravels may influence the use and management of land because of tillage difficulties but these larger particles make little or no contribution to soil properties such as WHC and capacity to store plant nutrients and their supply.

Gravels: 2 - 4 mm

Pebbles: 4 - 64 mm

Cobbles: 64 - 256 mm

Boulders: > 256 mm

Particles less than 2 mm are called fine earth, normally considered in chemical and mechanical analysis.

The Components of Fine Earth: Sand, Silt and Clay (Soil separates. The size limits of these fractions have been established by various organizations. There are a number of systems of naming soil separates.

1. (a) The American system developed by USDA
2. (b) The English system or British system (BSI)
3. (c) The International system (ISSS)
4. (d) European system

i) USDA

Soil separates	***Diameter in mm***
Clay	< 0.002
Silt	0.002 - 0.05
Very Fine Sand	0.05 - 0.10
Fine Sand	0.10 - 0.25
Medium Sand	0.25 - 0.50
Coarse Sand	0.50 - 1
Very Coarse Sand	1 - 2 mm

ii) BSI

Soil separates	***Diameter in mm***
Clay	< 0.002
Fine Silt	0.002 - 0.01
Medium Silt	0.01 - 0.04
Coarse Silt	0.04 - 0.06
Fine Sand	0.06 - 0.20
Medium Sand	0.20 - 1
Coarse Sand	1 - 2 mm

iii) ISSS

Soil separates	***Diameter in mm***
Clay	< 0.002
Silt	0.002 - 0.02
Fine Sand	0.02 - 0.2
Coarse Sand	0.2 - 2

iv) European System

Soil separates	***Diameter (mm)***
Fine clay	< 0.0002 mm
Medium clay	0.0002 – 0.0006
Coarse clay	0.0006 – 0.002
Fine silt	0.002 - 0.006
Medium silt	0.006 - 0.02
Coarse silt	0.02 - 0.06
Fine sand	0.06 - 0.20
Medium sand	0.20 - 0.60
Coarse sand	0. 60 - 2.00

Sand:

1. Usually consists of quartz but may also contain fragments of feldspar, mica and occasionally heavy minerals viz., zircon, Tourmaline and hornblende.
2. Has uniform dimensions
3. Can be represented as spherical
4. Not necessarily smooth and has jagged surface

Silt:

1. Particle size intermediate between sand and clay
2. Since the size is smaller, the surface area is more
3. Coated with clay
4. Has the physico- chemical properties as that of clay to a limited extent
5. Sand and Silt forms the SKELETON

Clay:

1. Particle size less than 0.002 mm
2. Plate like or needle like in shape

3. Belong to alumino silicate group of minerals
4. Some times considerable concentration of fine particles which does not belong to alumino silicates. (e.g.) iron oxide and CaCO3
5. These are secondary minerals derived from primary minerals in the rock
6. Flesh of the soil

Knowledge on Texture is important. It is a guide to the value of the land. Land use capability and methods of soil management depends on texture.

Particle Size Distribution/ Determination

The determination of relative distribution of the ultimate or individual soil particles below 2 mm diameter is called as Particle size analysis or Mechanical analysis

Two Steps are Involved

i) Separation of all the particles from each other ie. Complete dispersion into ultimate particles

ii) Measuring the amount of each group

Sr. No.	*Aggregating agents*	*Dispersion method*
1	Lime and Oxides of Fe & Al	Dissolving in HCl
2	Organic matter	Oxidises with H2O2
3	High conc. of electrolytes (soluble salts)	Precipitate and decant or filter with suction
4	Surface tension	Elimination of air by stirring with water or boiling

After removing the cementing agents, disperse by adding NaOH

Measurement

Once the soil particles are dispersed into ultimate particles, measurement can be done

i) Coarser fractions - sieving - sieves used in the mechanical analysis corresponds to the desired particle size separation for 2 mm, 1 mm and 0.5 mm – sieves with circular holes, for smaller sizes, wire mesh screens are used (screening)

ii) Finer fractions - by settling in a medium the settling or the velocity of the fall of particles is influenced by viscosity of the medium. Difference in density between the medium and falling particles, size and shape of object

Stokes' Law: Particle size analysis is based on a simple principle i.e. "when soil particles are suspended in water they tend to sink.

Because there is little variation in the density of most soil particles, their velocity (V) of settling is proportional to the square of the radius 'r' of each particle.

Thus V = kr2, where k is a constant. This equation is referred to as Stokes' law.

Stokes (1851) was the first to suggest the relationship between the radius of the particles and its rate of fall in a liquid. He stated that "the velocity of a falling particle is proportional to the square of the radius and not to its surface. The relation between the diameter of a particle and its settling velocity is governed by Stokes' Law:

V = 2/9 gr^2 (ds – dw) / n

Where,

V - Velocity of settling particle (cm/sec.)

g - Acceleration due to gravity cm/ sec2 (981)

ds - Density of soil particle (2.65)

dw - Density of water (1)

n - Coefficient of viscosity of water (0.0015 at 4oC)

r - Radius of spherical particles (cm).

Assumptions and Limitations of Stokes' Law

Particles are rigid and spherical / smooth. This requirement is very difficult to fulfill, because the particles are not completely smooth over the surface and spherical. It is established that the particles are not spherical and irregularly shaped such as plate and other shapes.

The particles are large in comparison with the molecules of the liquid so that in comparison with the particle the medium can be considered as homogenous i.e. the particles must be big enough to avoid Brownian movement. The particles less than 0.0002 mm exhibit this movement so that the rate of falling is varied.

The fall of the particles is not hindered or affected by the proximity (very near) of the wall of the vessel or of the adjacent particles. Many fast falling particles may drag finer particles down along with them.

The density of the particles and water and as well as the viscosity of the medium remain constant. But this is usually not so because of their different chemical and mineralogical composition.

The suspension must be still. Any movement in the suspension will alter the velocity of fall and such movement is brought by the sedimentation of larger particles (> 0.08 mm). They settle so fast and create turbulence in the medium. The temperature should be kept constant so that convection currents are not set up.

Methods of Textural Determination

Numerous methods for lab and field use have been developed

i) Elutriation method – Water & Air

ii) Pipette method

iii) Decantation/ beaker method

iv) Test tube shaking method

v) Feel method - Applicable to the field - quick method - by feeling the soil between thumb and fingers

Feel Method

Evaluated by attempting to squeeze the moistened soil into a thin ribbon as it is pressed with rolling motion between thumb and pre finger or alternately to roll the soil into a thin wire

η four aspects to be seen - i) Feel by fingers, ii) Ball formation, iii) Stickiness and iv) Ribbon formation

Soil Textural Classes

To convey an idea of the textural make up of soils and to give an indication of their physical properties, soil textural class names are used. These are grouped into three main fractions viz., Sand, Silt and Clay.

According to the proportion of these three fractions a soil is given a name to indicate its textural composition. Such a name gives an idea not only of the textural composition of a soil but also of its various properties in general.

On this basis soils are classified into various textural classes like sands clays, silts, loams etc

Sands: The sand group includes all soils in which the sand separates make up at least 70% and the clay separate 15% or less of the material by weight. The properties of such soils are therefore characteristically those of sand in contrast to the stickier nature of clays. Two specific textural classes are recognized in this group sandy and loamy sand although in practice two subclasses are also used Loamy fine sand and loamy very fine sand.

Silt: The silt group includes soils with at least 80% silt and 12% or less clay. Naturally the properties of this group are dominated by those of silt. Only one textural class - Silt is included in this group.

Clays: To be designated clay a soil must contain at least 35% of the clay separate and in most cases not less than 40%. In such soils the characteristics of the clay separates are distinctly dominant, and the class names are clay, sandy clay and silty clay. Sandy clays may contain more sand than clay. Likewise, the silt content of silty clays usually exceeds clay fraction.

Loams: The loam group, which contains many subdivisions, is a more complicated soil textural class. An ideal loam may be defined as a mixture of sand, silt and day particles that exhibits the properties of those separates in about equal proportions. Loam soils do not exhibit dominant physical properties of sand, silt or clay. Loam does not contain equal percentage of sand, silt and clay. However, exhibit approximately equal properties of sand, silt and clay.

Determination of Textural Class: In the American system as developed by the United State Department of Agriculture twelve textural classes are proposed.

The textural triangle: It is used to determine the soil textural name after the percentages of sand, silt, and clay are determined from a laboratory analysis. Since the soil's textural classification includes only mineral particles and those of less than 2mm diameter, the sand plus silt plus clay percentages equal 100 percent. (Note that organic matter is not included.) Knowing the amount of any two fractions automatically fixes the percentage of the third one.

To use the diagram, locate the percentage of clay first and project inward parallel to sand line. Do likewise for the per cent silt and project inward parallel to clay line and for sand, project inward parallel to silt. The point at which the projections cross or intersect will identify the class name.

Some times, the intersecting point exactly falls on the line between the textural classes. Then it is customary to use the name of the finer fraction when it happens. (e.g). Soil containing 40% clay, 30% sand and 30% silt - called as clay rather than clay loam.

Importance of Soil Texture

Presence of each type of soil particles makes its contribution to the nature and properties of soil as a whole

- Texture has good effect on management and productivity of soil. Sandy soils are of open character usually loose and friable.
- Such type of the texture is easy to handle in tillage operations.

- Sand facilitates drainage and aeration. It allows rapid evaporation and percolation.
- Sandy soils have very little water holding capacity. Such soils can not stand drought and unsuitable for dry farming.
- Sandy soils are poor store house of plant nutrients
- Contain low organic matter
- Leaching of applied nutrients is very high.
- In sandy soil, few crops can be grown such as potato, groundnut and cucumbers.
- Clay particles play a very important role in soil fertility.
- Clayey soils are difficult to till and require much skill in handling. When moist clayey soils are exceedingly sticky and when dry, become very hard and difficult to break.
- They have fine pores, and are poor in drainage and aeration.
- They have a high water holding capacity and poor percolation, which usually results in water logging.
- They are generally very fertile soils, in respect of plant nutrient content. Rice, jute, sugarcane can be grown very successfully in these soils.
- Loam and Silt loam soils are highly desirable for cultivation
- Generally, the best agriculture soils are those contain 10 – 20 per cent clay, 5 – 10 per cent organic matter and the rest equally shared by silt and sand

Soil Structure

Soil conditions and characteristics such as water movement, heat transfer, aeration, and porosity are much influenced by structure. In fact, the important physical changes imposed by the farmer in ploughing, cultivating, draining, liming, and manuring his land are structural rather than textural.

Definition of Soil Structure: The arrangement and organization of primary and secondary particles in a soil mass is known as soil structure.

Soil structure controls the amount of water and air present in soil. Plant roots and germinating seeds require sufficient air and oxygen for respiration.

Bacterial activities also depend upon the supply of water and air in the soil.

Formation of soil structure: Soil particles may be present either as single individual grains or as aggregate i.e. group of particles bound together into granules or compound particles. These granules or compound particles are known as secondary particles. A majority of particles in a sandy or silty soil are present as single individual grains while in clayey soil they are present in granulated condition. The individual particles are usually solid, while the aggregates are not solid but they possess a porous or spongy character. Most soils are mixture of single grain and compound particle. Soils, which predominate with single grains are said to be structure less, while those possess majority of secondary particles are said to be aggregate, granulated or crumb structure.

Mechanism of Aggregate Formation: The bonding of the soil particles into structural unit is the genesis of soil structure. The bonding between individual particles in the structural units is generally considered to be stronger than the structural units themselves.

In aggregate formation, a number of primary particles such as sand, silt and clay are brought together by the cementing or binding effect of soil colloids. The cementing materials taking part in aggregate formation are colloidal clay, iron and aluminium hydroxides and decomposing organic matter. Whatever may be the cementing material, it is ultimately the dehydration of colloidal matter accompanied with pressure that completes the process of aggregation.

Colloidal clay: By virtue of high surface area and surface charge, clay particles play a key role in the formation of soil aggregates. Sand and silt particles can not form aggregates as they do not possess the power of adhesion and cohesion. These particles usually carry a coating of clay particles; they are enmeshed in the aggregates formed by the adhering clay particles. Colloidal particles form aggregates only when they are flocculated. There is vast difference between flocculation and aggregation.

Flocculation is brought about by coalescence of colloidal particles and is the first step in aggregation.

Aggregation is some thing more than flocculation involving a combination of different factors such as hydration, pressure, dehydration etc. and required cementation of flocculated particles. The cementation may be caused by cations, oxides of Fe and Al, humus substances and products of microbial excretion and synthesis. Clay particles form aggregates only if they are wetted by a liquid like water whose molecules possess an appreciable dipole moment.

Clay - - +Water - - +Cation+ - -Clay - - +Water - - +Cation+ - - Clay -

The aggregation also depends upon the nature of clay particles, size and amount of clay particles, dehydration of clay particles, cations like calcium and anions like phosphate.

Fe and Al oxides: The colloidal Fe oxides act as cementing agent in aggregation. Al oxides bind the sand and silt particles. These act in two ways. A part of the hydroxides acts as a flocculating agent and the rest as a cementing agent.

Organic matter: It also plays an important role in forming soil aggregates.

1. During decomposition, cellulose substances produce a sticky material very much resembling mucus or mucilage. The sticky properly may be due to the presence of humic or humic acid or related compounds produced.
2. Certain polysaccharides formed during decomposition.
3. Some fungi and bacteria have cementing effect probably due to the presence of slimes and gums on the surface of the living organisms produced as a result of the microbial activity

Classification of Soil Structure: The primary particles sand, silt and clay usually occur grouped together in the form of aggregates.

Natural aggregates are called peds where as clod is an artificially formed soil mass. Structure is studied in the field under natural conditions and it is described under three categories

1. Type - Shape or form and arrangement pattern of peds
2. Class - Size of Peds
3. Grade - Degree of distinctness of peds

Types of Soil Structure: There are four principal forms of soil structure

Plate-like (Platy): In this type, the aggregates are arranged in relatively thin horizontal plates or leaflets. The horizontal axis or dimensions are larger than the vertical axis. When the units/ layers are thick they are called platy. When they are thin then it is laminar.

Platy structure is most noticeable in the surface layers of virgin soils but may be present in the subsoil.

This type is inherited from the parent material, especially by the action of water or ice.

Prism-like: The vertical axis is more developed than horizontal, giving a pillar like shape. Vary in length from 1- 10 cm. commonly occur in sub soil horizons of Arid and Semi arid regions. When the tops are rounded, the structure is termed as columnar when the tops are flat / plane, level and clear cut prismatic.

Block like: All three dimensions are about the same size. The aggregates have been reduced to blocks. Irregularly six faced with their three dimensions more or less equal.

When the faces are flat and distinct and the edges are sharp angular, the structure is named as angular blocky. When the faces and edges are mainly rounded it is called sub angular blocky. These types usually are confined to the sub soil and characteristics have much to do with soil drainage, aeration and root penetration.

Spheroidal (Sphere like): All rounded aggregates (peds) may be placed in this category. Not exceeding an inch in diameter. These rounded complexes usually loosely arranged and readily separated. When wetted, the intervening spaces generally are not closed so readily by swelling as may be the case with a blocky structural condition.

Therefore in sphere like structure, infiltration, percolation and aeration are not affected by wetting of soil.

The aggregates of this group are usually termed as granular which are relatively less porous. When the granules are very porous, it is termed as crumb. This is specific to surface soil particularly high in organic matter/ grass land soils.

Classes of Soil Structure: Each primary structural type of soil is differentiated into 5 size classes depending upon the size of the individual peds.

The terms commonly used for the size classes are:

1. Very fine or very thin
2. Fine or thin
3. Medium
4. Coarse or thick
5. Very Coarse or very thick

The terms thin and thick are used for platy types, while the terms fine and coarse are used for other structural types.

Grades of Soil Structure: Grades indicate the degree of distinctness of the individual peds. It is determined by the stability of the aggregates.

Grade of structure is influenced by the moisture content of the soil. Grade also depends on organic matter, texture etc. Four terms commonly used to describe the grade of soil structure are:

1. Structure less: There is no noticeable aggregation, such as conditions exhibited by loose sand.
2. Weak Structure: Poorly formed, indistinct formation of peds, which are not durable and much unaggregated material.
3. Moderate structure: Moderately well developed peds, which are fairly durable and distinct.
4. Strong structure: Very well formed peds, which are quite durable and distinct.

Soil Structure Naming: For naming a soil structure the sequence followed is grade, class and type; for example strong coarse angular blocky, moderate thin platy, weak fine prismatic.

Factors Affecting Soil Structure

The development of structure in arable soil depends on the following factors:

1. Climate: Climate has considerable influence on the degree of aggregation as well as on the type of structure. In arid regions there is very little aggregation of primary particles. In semi arid regions, the degree of aggregation is greater.
2. Organic matter: Organic matter improves the structure of a sandy soil as well as of a clay soil. In case of a sandy soil, the sticky and slimy material produced by the decomposing organic matter and the associated microorganism cement the sand particles together to form aggregates. In case of clayey soil, it modifies the properties of clay by reducing its cohesiveness. This helps making clay more crumby.
3. Tillage: Cultivation implements break down the large clods into smaller fragments and aggregates. For obtaining good granular and crumby structure, optimum moisture content in the soil is necessary. If the moisture content is too high it will form large clods on drying. If it is too low some of the existing aggregates will be broken down.
4. Plants, Roots and Residues: Excretion of gelatinous organic compounds and exudates from roots serve as a link. Root hairs make soil particles to cling together. Grass and cereal roots vs other roots. Pressure exerted by the roots also held the particles together.

Dehydration of soil strains the soil due to shrinkage result in cracks lead to aggregation

Plant tops and residues shade the soil prevent it from extreme and sudden temperature and moisture changes and also from rain drop impedance.

Plant residues serve as a food to microbes which are the prime aggregate builders.

5. Animals: Among the soil fauna small animals like earthworms, moles and insects etc., that burrow in the soil are the chief agents that take part in the aggregation of finer particles.
6. Microbes: Algae, fungi, actinomycetes and fungi keep the soil particles together. Fungi and actinomycetes exert mechanical binding by mycelia, Cementation by the products of decomposition and materials synthesized by bacteria.
7. Fertilizers: Fertilizer like Sodium Nitrate destroys granulation by reducing the stability of aggregates. Few fertilizers for example, CAN help in development of good structures.
8. Wetting and drying: When a dry soil is wetted, the soil colloids swell on absorbing water. On drying, shrinkage produces strains in the soil mass gives rise to cracks, which break it up into clods and granules of various sizes.
9. Exchangeable cations: Ca, Mg ————à H, Na

   Flocculating Deflocculating

   Good structure Poor structure
10. Inorganic cements: CaCO3 and Sesquioxides
11. Clay
12. Water

Effect of Soil Structure on other Physical Properties

Porosity: Porosity of a soil is easily changed. In plate like structure, pore spaces are less where as in crumby structure pore spaces are more.

Temperature: Crumby structure provides good aeration and percolation of water in the soil. Thus these characteristics help in keeping optimum temperature in comparison to plate like structure.

Density: Bulk density varies with the total pore space present in the soil. Structure chiefly influences pore spaces Platy structure with less total pore spaces has high bulk density where as crumby structure with more total pore spaces has low bulk density.

Consistence: Consistence of soil also depends on structure. Plate-like structure exhibits strong plasticity.

Colour: Bluish and greenish colours of soil are generally due to poor drainage of soil. Platy structure normally hinders free drainage.

Importance of Structure: Soil structure influences rather indirectly by the formation of an array of pores of various shapes and sizes. These pores are controlling factors governing water, air and temperature in soil.

Role of Soil Structure in Relation to Plant Growth

1. Soil structure influences the amount and nature of porosity.
2. Structure controls the amount of water and air present in the soil. Not only the amount of water and air dependent on soil structure, but their movement and circulation are also controlled by soil structure.
3. It affects tillage practices.
4. Structure controls runoff and erosion.
5. Platy structure normally hinders free drainage whereas sphere like structure (granular and crumby) helps in drainage.
6. Crumby and granular structure provides optimum infiltration, water holding capacity, aeration and drainage. It also provides good habitat for microorganisms and supply of nutrients.

Class of Soil Structure as differentiated by size of soil peds

Class	*Platy*	*Prism-atic*	*Colum-nar*	*Blocky*	*S.A. Blocky*	*Gran-ular*	*Crumb*
V.Fine or V.Thin	<1	<10	<10	<5	<5	<1	<1
Fine or Thin	1-2	10-20	10-20	5-10	5-10	1-2	1-2
Medi	2-5	20-50	20-50	10-20	10-20	2-5	2-5
um	5-10	50-100	50-100	20-50	20-50	5-10	-
Coarseor Thick	>10	>100	>100	>50	>50	>10	-
V.C or V.Thic	>10	>100	>100	>50	>50	>10	-

Density of Soil: Bulk Density and Particle Density

Density represents weight (mass) per unit volume of a substance.

Density = Mass / Volume

Soil density is expressed in two well accepted concepts as particle density and bulk density. In the metric system, particle density can be expressed in terms of mega grams per cubic centre (Mg/m3). Thus if 1 m3 of soil solids weighs 2.6 Mg, the particle density is 2.6 Mg / m3 (since 1 Mg =1 million grams and 1 m3 =1 million cubic centimeters) thus particle density can also be expressed as 2.6 g / cm3.

Particle Density: The weight per unit volume of the solid portion of soil is called particle density. Generally particle density of normal soils is 2.65 grams per cubic centimeter. The particle density is higher if large amount of heavy minerals such as magnetite; limonite and hematite are present in the soil. With increase in organic matter of the soil the particle density decreases. Particle density is also termed as true density.

Table: *Particle density of different soil textural classes*

Textural classes	***Particle density (g/ cm3)***
Coarse sand	2.655
Fine sand	2.659
Silt	2.798
Clay	2.837

Bulk Density: The oven dry weight of a unit volume of soil inclusive of pore spaces is called bulk density. The bulk density of a soil is always smaller than its particle density. The bulk density of sandy soil is about 1.6 g / cm3, whereas that of organic matter is about 0.5. Bulk density normally decreases, as mineral soils become finer in texture. The bulk density varies indirectly with the total pore space present in the soil and gives a good estimate of the porosity of the soil. Bulk density is of greater importance than particle density in understanding the physical behaviour of the soil. Generally soils with low bulk densities have favourable physical conditions.

Table: *Bulk density of different textural classes*

Textural class	***Bulk density (g/cc)***	***Pore space (%)***
Sandy soil	1.6	40
Loam	1.4	47
Silt loam	1.3	50
Clay	1.1	58

Factors Affecting Bulk Density

1. Pore space: Since bulk density relates to the combined volume of the solids and pore spaces, soils with high proportion of pore space to solids have lower bulk densities than those that are more compact and have less pore space. Consequently, any factor that influences soil pore space will affect bulk density.
2. Texture: Fine textured surface soils such as silt loams, clays and clay loams generally have lower bulk densities than sandy soils. This is because the fine textured soils tend to organize in porous grains especially because of adequate organic matter content. This results in high pore space and low bulk density. However, in sandy soils, organic matter content is generally low, the solid particles lie close together and the bulk density is commonly higher than in fine textured soils.
3. Organic matter content: More the organic matter content in soil results in high pore space there by shows lower bulk density of soil and vice-versa.

Soil Colloids

The colloidal state refers to a two-phase system in which one material in a very finely divided state is dispersed through second phase.

The Examples Are: Solid in liquid - Clay in water (dispersion of clay in water)

Liquid in gas -Fog or clouds in atmosphere

The clay fraction of the soil contains particles less than 0.002 mm in size. Particles less than 0.001 mm size possess colloidal properties and are known as soil colloids.

General Properties of Soil Colloids

1. Size: The most important common property of inorganic and organic colloids is their extremely small size. They are too small to be seen with an ordinary light microscope. Only with an electron microscope they can be seen. Most are smaller than 2 micrometers in diameter.
2. Surface area: Because of their small size, all soil colloids expose a large external surface per unit mass. The external surface area of 1 g of colloidal clay is at least 1000 times that of 1 g of coarse sand. Some colloids, especially certain silicate clays have extensive internal surfaces as well. These internal surfaces

occur between plate like crystal units that make up each particle and often greatly exceed the external surface area. The total surface area of soil colloids ranges from 10 m2/g for clays with only external surfaces to more than 800 m2/g for clays with extensive internal surfaces. The colloid surface area in the upper 15 cm of a hectare of a clay soil could be as high 700,000 km2/g

3. Surface charges: Soil colloidal surfaces, both external and internal characteristically carry negative and/or positive charges. For most soil colloids, electro negative charges predominate. Soil colloids both organic and inorganic when suspended in water, carry a negative electric charge. When an electric current is passed through a suspension of soil colloidal particles they migrate to anode, the positive electrode indicating that they carry a negative charge. The magnitude of the charge is known as zeta potential. The presence and intensity of the particle charge influence the attraction and repulsion of the particles towards each other, there by influencing both physical and chemical properties.

   a) Ionizable hydrogen ions and

   b) Isomorphism substitution.

      i) Ionizable hydrogen ions: Ionizable hydrogen ions are hydrogen from hydroxyl ions on clay surfaces. The -Al-OH or -Si-OH portion of the clay ionizes the H and leaves an unneutralized negative charge on the oxygen (-Al-O- or - Si-O). The extent of ionized hydrogen depends on solution pH; more ionization occurs in more alkaline (basic) solutions.

      ii) Isomorphous substitution: The second source of charge on clay particles is due to the substitution of one ion for another of similar size and often with lower positive valence. In clay structures, certain ions fit into certain mineral lattice sites because of their convenient size and charge. Dominantly, clays have Si4+ in tetrahedral sites and A13+ in octahedral sites. Other ions present in large amounts during clay crystallization can replace some of the A13+ and Si4+ cations. Substitutions that are common are the Si4+ replaced by A13+, and even more extensive replacement of A13+ by one or more of these: Fe3+, Fe2+, Mg2+ or Zn2+ Since the total negative charge from the anions (the oxygen) remains unchanged, the lower positive charge because of substitution results in an excess negative charge at that location in the structure.

4. Adsorption of cations: As soil colloids possess negative charge they attract the ions of an opposite charge to the colloidal surfaces. They attract hundreds of positively charged ions or cation such as H+, A13+ Ca2+ , and Mg2+. This gives rise to an ionic double layer.

   The process, called Isomorphous substitution and the colloidal particle constitutes the inner ionic layer, being essentially huge anions; with both, external and internal layers that are negative in charge. The outer layer is made up of a swarm of rather loosely held (adsorbed) cations attracted to the negatively charged surfaces. Thus a colloidal particle is accompanied by a swarm of cations that are adsorbed or held on the particle surfaces.

5. Adsorption of water: In addition to the adsorbed cations, a large number of water molecules are associated with soil colloidal particles. Some are attracted to the adsorbed cations, each of which is hydrated; others are held in the internal surfaces of the colloidal particles. These water molecules play a critical role in determining both the physical and chemical properties of soil.

6. Cohesion: Cohesion is the phenomenon of sticking together of colloidal particles that are of similar nature. Cohesion indicates the tendency of clay particles to stick together. This tendency is primarily due to the attraction of the clay particles for the water molecules held between them. When colloidal substances are wetted, water first adheres to the particles and then brings about cohesion between two or more adjacent colloidal particles.

7. Adhesion: Adhesion refers to the phenomenon of colloidal particles sticking to other substances. It is the sticking of colloida1 materials to the surface of any other body or substance with which it comes in contact.

8. Swelling and shrinkage: Some clay (soil colloids) such as smectites swell when wet and shrink when dry. After a prolonged dry spell, soils high in smectites (e.g. Vertisols) often are crises-crossed by wide, deep cracks, which at first allow rain to penetrate rapidly. Later, because of swelling, such soil

is likely to close up and become much more impervious than one dominated by kaolinite, chlorite, or fine grained micas. Vermiculite is intermediate in its swelling and shrinking characteristics.

9. Dispersion and flocculation: As long as the colloidal particles remain charged, they repel each other and the suspension remains stable. If on any account they loose their charge, or if the magnitude of the charge is reduced, the particles coalesce, form flocs or loose aggregates, and settle out. This phenomenon of coalescence and formation of flocs is known as flocculation. The reverse process of the breaking up of flocs into individual particles is known as deflocculation or dispersion.
10. Brownian movement: When a suspension of colloidal particles is examined under a microscope the particles seem to oscillate. The oscillation is due to the collision of colloidal particles or molecules with those of the liquid in which they are suspended. Soil colloidal particles with those of water in which they are suspended are always in a constant state of motion. The smaller the particle, the more rapid is its movement.
11. Non permeability: Colloids, as opposed to crystalloid, are unable to pass through a semi-permeable membrane. Even though the colloidal particles are extremely small, they are bigger than molecules of crystalloid dissolved in water. The membrane allows the passage of water and of the dissolved substance through its pores, but retains the colloidal particles.

Types of Soil Colloids

There are four major types of colloids present in soil

1. Layer silicate clays
2. Iron and aluminum oxide clays (sesquioxide clays)
3. Allophane and associated amorphous clays
4. Humus.

Layer silicate clays, iron and aluminum oxide clays, allophane and associated amorphous clays are inorganic colloids while humus is an organic colloid.

Layer Silicate Clays

These are most important silicate clays and are known as phyllosilicates (Phyllon - leaf) because of their leaf-like or plate like structure. They are comprised of two kinds of horizontal sheets. One dominated by silicon and other by aluminum and/or magnesium.

Silica Tetrahedron: The basic building block for the silica-dominated sheet is a unit composed of one silicon atom surrounded by four oxygen atoms. It is called the silica tetrahedron because of its four-sided configuration. An interlocking array or a series of these silica tetrahedral tied together horizontally by shared oxygen anions gives a tetrahedral sheet.

Alumina Octahedron: Aluminium and/or magnesium ions are the key cations in the second type of sheet. An aluminium (or magnesium) ion surrounded by six oxygen atoms or hydroxyl group gives an eight sided building block termed octahedron. Numerous octahedra linked together horizontally comprise the octahedral sheet. An aluminum dominated sheet is known as a dioctahedral sheet, whereas one dominated by magnesium is called a trioctahedral sheet. The distinction is due to the fact that two aluminum ions in a dioctahedral sheet satisfy the same negative charge from surrounding oxygen and hydroxyls as three magnesium ions in a trioctahedral sheet.

The tetrahedral and octahedral sheets are the fundamental structural units of silicate clays. They, in turn, are bound together within the crystals by shared oxygen atoms into different layers. The specific nature and combination of sheets in these layers vary from one type of clay to another and largely control the physical and chemical properties of each clay.

Types of Silicate Clay Minerals: On the basis of the number and arrangement of tetrahedral (silica) and octahedral (alumina-magnesia) sheets contained in the crystal units or layers, silicate clays are classified into three different groups

A. 1 :1 Type clay minerals

B. 2:1 Type clay minerals

C. 2: 1: 1 Type clay minerals

A. 1:1 Type Minerals: The layers of the 1:1-type minerals are made up of one tetrahedral (silica) sheet combined with one octahedral (alumina) sheet-hence the terminology. In soils, kaolinite is the most prominent member of this group, which includes hallosite, nacrite, and dickite.

The tetrahedral and octahedral sheets in a layer of a kaolinite crystal are held together tightly by oxygen atoms, which are mutually shared by the silicon and aluminum cations in their respective sheets. These layers, in turn, are held together by hydrogen bonding. Consequently, the structure is fixed and no expansion ordinarily occurs between layers when the clay is wetted.

Cations and water do not enter between the structural layers of a 1:1 type mineral particle. The effective surface of kaolinite is thus restricted to its outer faces or to its external surface area. Also, there is little isomorphous substitution in this 1:1 type mineral. Along with the relatively low surface area of kaolinite, this accounts for its low capacity to adsorb cations.

Kaolinite crystals usually are hexagonal in shape. In comparison with other clay particles, they are large in size, ranging from 0.10 to 5 um across with the majority falling within the 0.2 to 2 um range. Because of the strong binding forces between their structural layers, kaolinite particles are not readily broken down into extremely thin plates.

Kaolinite exhibits very little plasticity (capability of being molded), cohesion, shrinkage, and swelling.

2:1-Type Minerals: The crystal units (layers) of these minerals are characterized by an octahedral sheet sandwiched between two tetrahedral sheets. Three general groups have this basic crystal structure.

i) Expanding type: smectites and vermiculite

ii) Non-expanding type: mica (illite)

Expanding Minerals: The smectite group is noted for interlayer expansion, which occurs by swelling when the minerals are wetted, the water entering the interlayer space and forcing the layers apart. Montmorillonite is the most prominent member of this group in soils, although beidellite, nontronite, and saponite are also found.

The flake-like crystals of smectites (e.g., Montmorillonite) are composed of an expanding lattice 2:1-type clay mineral. Each layer is made up of an octahedral sheet sandwiched between two tetrahedral (silica) sheets. There is little attraction between oxygen atoms in the bottom tetrahedral sheet of one unit and those in the top tetrahedral sheet of another. This permits a ready and variable space between layers, which is occupied by water and exchangeable cations. This internal surface far exceeds the surface around the outside of the crystal. In montmorillonite magnesium has replaced aluminum in some sites of the octahedral sheet. Likewise, some silicon atoms in the tetrahedral sheet may be replaced by aluminum. These substitutions give rise to a negative charge.

These minerals show high cation exchange capacity, marked swelling and shrinkage properties. Wide cracks commonly form as smectite dominated soils (e.g., Vertisols) are dried. The dry aggregates or clods are very hard, making such soils difficult to till.

Vermiculites are also 2: 1 type minerals in that an octahedral sheet occurs between two tetrahedral sheets. In most soils vermiculites, the octahedral sheet is aluminum dominated (dioctahedral), although magnesium dominated (trioctahedral) vermiculites are also common. In the tetrahedral sheet of most vermiculite, considerable substitution of aluminum for silicon has taken place. This accounts for most of the very high net negative charge associated with these minerals.

Water molecules, along with magnesium and other ions, are strongly adsorbed in the interlayer space of vermiculites. They act primarily as bridges holding the units together rather than as wedges driving them apart. The degree of swelling is, therefore considerable less for vermiculites than for smectites. For this reason, vermiculites are considered limited-expansion clay minerals, expanding more than kaolinite but much less than the smectites.

The cation exchange capacity of vermiculites usually exceeds that of all other silicate clays, including montmorillonite and other smectites, because of very high negative charge in the tetrahedral sheet. Vermiculite crystals are larger than those of the smectites but much smaller than those of kaolinite.

Non-expanding Minerals: Micas are the type minerals in this group. (e.g.) Muscovite and biotite, weathered minerals similar in structure to these micas are found in the clay fraction of soils. They are called fine-grained micas. Like sanctities, fine-grained micas have a 2:1-type crystal. However, the particles are much larger than those of the smectites. Also, the major source of charge is in-the tetrahedral sheet where aluminum atoms occupy about 20% of the silicon sites. These results in a high net negative charge in the tetrahedral sheet, even higher than that found in vermiculites, to satisfy this charge, potassium ions are strongly attracted in the interlayer space and are just the right size to fit into certain spaces in the adjoining tetrahedral sheets. The potassium thereby acts as a binding agent, preventing expansion of the crystal. Hence, fine-grained micas are quite non expansive.

The properties such as hydration, cation adsorption, swelling, shrinkage, and plasticity are much less intense in fine-grained micas than in smectites. The fine grained micas exceed kaolinite with respect to these characteristics, but this may be due in part to the presence of interstratified layers of smectite or vermiculite. In size, too, fine-grained mica crystals are intermediate between the smectities and kaolihites.

Their specific surface area varies from 70 to 100 m2/g, about one eighth that for the smectites.

2:1:1 Type Minerals: This silicate group is represented by chlorites, which are common in a variety of soils. Chlorites are basically iron magnesium silicates with some aluminum present. In a typical chlorite clay crystal, 2:1 layers, such as in vermiculites, alternate with a magnesium-dominated trioctahedral sheet, giving a 2:1:1 ratio. Magnesium also dominates the trioctahedral sheet in the 2:1 layer of chlorites. Thus, the crystal unit contains two silica tetrahedral sheets and two magnesium-dominated trioctahedral sheets giving rise to the term 2:1:1 or 2:2-type structure.

The negative charge of chlorites is about the same as that of fine-grained micas considerably less than that of the smectites or vermiculites. Like fine micas, chlorites may be interstratified with vermiculites or smectites in a single crystal. Particle size and surface area for chlorites are also about the same as for fine grained micas. There is no water adsorption between the chlorite crystal units, which accounts for the non expanding nature of this mineral.

Mixed and interstratified layers: Specific groups of clay minerals do not occur independently of one another. In a given soil, it is common to find several clay minerals in an intimate mixture. Furthermore, some mineral colloids have properties and composition intermediate between those of any two of the well defined minerals described. Such minerals are termed mixed layer or interstratified because the individual layers within a given crystal may be of more than one type. Terms such as "chlorite-vermiculite" and "fine-grained mica- smectite" are used to describe mixed-layer minerals. In some soils, they are more common than single-structured minerals such as montmorillonite.

Iron and Aluminum Oxide Clays (Sesquioxide Clays): Under conditions of extensive leaching by rainfall and long time intensive weathering of minerals in humid warm climates, most of the silica and much of the alumina in primary minerals are dissolved and slowly leached away. The remnant materials, which have lower solubility, are sesquioxides. Sesquioxides (metal oxides) are mixtures of aluminum hydroxide, Al $(OH)_3$, and iron oxide, Fe_2O_3, or iron hydroxide, Fe $(OH)_3$.

The Latin word sesqui means one and one- half times, meaning one and one-half times more oxygen than Al and Fe. These clays can grade from amorphous to crystalline.

Examples of iron and aluminum oxides common in soils are gibbsite ($Al_2O_3.3H_2O$) and geothite ($Fe_2O_3.H_2O$).

Less is known about these clays than about the layer silicates. These clays do not swell, not sticky and have high phosphorus adsorption capacity

Allophane and Other Amorphous Minerals: These silicate clays are mixtures of silica and alumina. They are amorphous in nature. Even mixture of other weathered oxides (iron oxide) may be a part of the mixture. Typically, these clays occur where large amount of weathered products existed. These clays are common in soils forming from volcanic ash (e.g., Allophane). These clays have high anion exchange capacity or even high cation exchange capacity. Almost all of their charge is from accessible hydroxyl ions (OH^-), which can attract a positive ion or lose the H^+ attached. These clays have a variable charge that depends on H^+ in solution (the soil acidity).

Humus (Organic Colloid): Humus is amorphous, dark brown to black, nearly insoluble in water, but mostly soluble in dilute alkali (NaOH or KOH) solutions. It is a temporary intermediate product left after considerable decomposition of plant and animal remains. They are temporary intermediate because the organic substances remain continue to decompose slowly.

The humus is often referred to as an organic colloid and consists of various chains and loops of linked carbon atoms. The humus colloids are not crystalline. They are composed basically of carbon, hydrogen, and oxygen rather than of silicon, aluminum, iron, oxygen, and hydroxyl groups.

The organic colloidal particles vary in size, but they may be at least as small as the silicate clay particles. The negative charges of humus are associated with partially dissociated enolic (-OH), carboxyl (-COOH), and phenolic groups; these groups in turn are associated with central units of varying size and complexity.

8

Soil Moisture

Water contained in soil is called soil moisture. The water is held within the soil pores. Soil water is the major component of the soil in relation to plant growth. If the moisture content of a soil is optimum for plant growth, plants can readily absorb soil water. Not all the water, held in soil, is available to plants. Much of water remains in the soil as a thin film. Soil water dissolves salts and makes up the soil solution, which is important as medium for supply of nutrients to growing plants.

Importance of Soil Water

1. Soil water serves as a solvent and carrier of food nutrients for plant growth
2. Yield of crop is more often determined by the amount of water available rather than the deficiency of other food nutrients
3. Soil water acts as a nutrient itself
4. Soil water regulates soil temperature
5. Soil forming processes and weathering depend on water
6. Microorganisms require water for their metabolic activities
7. Soil water helps in chemical and biological activities of soil
8. It is a principal constituent of the growing plant
9. Water is essential for photosynthesis

Retention of Water by Soil: The soils hold water (moisture) due to their colloidal properties and aggregation qualities. The water is held on the surface of the colloids and other particles and in the pores. The forces responsible for retention of water in the soil after the drainage has stopped are due to surface tension and surface attraction and are called surface moisture tension. This refers to the energy concept in moisture retention relationships. The force with which water is held is also termed as suction.

Ways to Retain Water in Soil

1. Cohesion and adhesion forces: These two basic forces are responsible for water retention in the soil. One is the attraction of molecules for each other i.e., cohesion. The other is the attraction of water molecules for the solid surface of soil i.e. adhesion. By adhesion, solids (soil) hold water molecules rigidly at their soil - water interfaces. These water molecules in turn hold by cohesion. Together, these forces make it possible for the soil solids to retain water.
2. Surface tension: This phenomenon is commonly evidenced at water- air interfaces. Water behaves as if its surface is covered with a stretched elastic membrane. At the surface, the attraction of the air for the water molecules is much less than that of water molecules for each other. Consequently, there is a net downward force on the surface molecules, resulting in sort of a compressed film (membrane) at the surface. This phenomenon is called surface tension.
3. Polarity or dipole character: The retention of water molecules on the surface of clay micelle is based on the dipole character of the molecule of water. The water molecules are held by electrostatic force that exists on the surface of colloidal particles. By virtue of their dipole character and under the influence of electrostatic forces, the molecules of water get oriented (arranged) on the surface of the clay particles in a particular manner.

Each water molecule carries both negative and positive charges. The clay particle is negatively charged. The positive end of water molecule gets attached to the negatively charged surface of clay and leaving its negative end outward. The water molecules attached to the clay surface in this way present a layer of negative charges to which another layer of oriented water molecules is attached. The number of successive molecular layers goes on increasing as long as the water molecules oriented. As the molecular layer gets thicker, orientation becomes weaker, and at a certain distance from the particle surface the water molecules cease to orientate and capillary water (liquid water) begins to appear. Due to the forces of adsorption (attraction) exerted by the surface of soil particles, water gets attached on the soil surface. The force of gravity also acts simultaneously, which tries to pull it downwards. The surface force is far greater than the force of gravity so water may remain attached to the soil particle. The water remains attached to the soil particle or move downward into the lower layers, depending on the magnitude of the resultant force.

Factors Affecting Soil Water

1. Texture: Finer the texture, more is the pore space and also surface area, greater is the retention of water.
2. Structure: Well-aggregated porous structure favours better porosity, which in turn enhance water retention.
3. Organic matter: Higher the organic matter more is the water retention in the soil.
4. Density of soil: Higher the density of soil, lower is the moisture content.
5. Temperature: Cooler the temperature, higher is the moisture retention.
6. Salt content: More the salt content in the soil less is the water available to the plant.
7. Depth of soil: More the depth of soil more is the water available to the plant.
8. Type of clay: The 2:1 type of day increases the water retention in the soil.

Soil Water Potential

The retention and movement of water in soils, its uptake and translocation in plants and its loss to the atmosphere are all energy related phenomenon. The more strongly water is held in the soil the greater is the heat (energy) required. In other words, if water is to be removed from a moist soil, work has to be done against adsorptive forces. Conversely, when water is adsorbed by the soil, a negative amount of work is done. The movement is from a zone where the free energy of water is high (standing water table} to one where the free energy is low (a dry soil). This is called soil water energy concept.

Free energy of soil solids for water is affected by

i) Matric (solid) force i.e., the attraction of the soil solids for water (adsorption} which markedly reduces the free energy (movement} of the adsorbed water molecules.

ii) Osmotic force i.e., the attraction of ions and other solutes for water to reduce the free energy of soil solution.

Matric and Osmotic potentials are negative and reduce the free energy level of the soil water. These negative potentials are referred as suction or tension.

iii) Force of gravity: This acts on soil water, the attraction is towards the earth's centre, which tends to pull the water down ward. This force is always positive. The difference between the energy states of soil water and pure free water is known as soil water potential. Total water potential (Pt} is the sum of the contributions of gravitational potential (Pg), matric potential (Pm) and the Osmotic potential or solute potential (Po).

Pt = Pg + Pm + Po

Potential represents the difference in free energy levels of pure water and of soil water. The soil water is affected by the force of gravity, presence of soil solid (matric) and of solutes.

Methods of expressing suctions: There are two units to express differences in energy levels of soil water.

i) PF Scale: The free energy is measured in terms of the height of a column of water required to produce necessary suction or pressure difference at a particular soil moisture level. The pF, therefore, represents the logarithm of the height of water column (cm) to give the necessary suction.

ii) Atmospheres or Bars: It is another common mean of expressing suction. Atmosphere is the average air pressure at sea level. If the suction is very low as occurs in the case of a wet soil containing the maximum amount of water that it can hold, the pressure difference is of the order of about 0.01 atmospheres or 1 PF equivalent to a column of water 10 cm in height. Similarly, if the pressure difference is 0.1 atmosphere the PF will be 20. Soil moisture constants can be expressed in term of PF values. A soil that is saturated with water has PF 0 while an oven dry soil has a PF 7.

Measuring Soil Moisture

Two general types of measurements relating to soil water are ordinarily used

i) By some methods the moisture content is measured directly or indirectly

ii) Techniques are used to determine the soil moisture potential (tension or suction)

Measuring soil moisture content in laboratory:

1. Gravimetric method: This consists of obtaining a moist sample, drying it in an oven at 105°C until it losses no more weight and then determining the percentage of moisture. The gravimetric method is time consuming and involves laborious processes of sampling, weighing and drying in laboratory.
2. Electrical conductivity method: This method is based upon the changes in electrical conductivity with changes in soil moisture. Gypsum blocks inside of with two electrodes at a definite distance are apart used in this method. These blocks require previous calibration for uniformity. The blocks are buried in the soil at desired depths and the conductivity across the electrodes measured with a modified Wheatstone bridge. These electrical measurements are affected by salt concentration in the soil solution and are not very helpful in soils with high salt contents.
3. Suction method or equilibrium tension method: Field tensiometers measure the tension with which water is held in the soils. They are used in determining the need for irrigation. The tensiometer is a porous cup attached to a glass tube, which is connected to a mercury monometer. The tube and cup are filled with water and cup inserted in the soil. The water flows through the porous cup into the soil until equilibrium is established. These tension readings in monometer, expressed in terms of cm or atmosphere, measures the tension or suction of the soil.

If the soil is dry, water moves through the porous cup, setting up a negative tension (or greater is the suction). The tensiometers are more useful in sandy soils than in fine textured soils. Once the air gets entrapped in the tensiometer, the reliability of readings is questionable.

Classification of Soil Water

Soil water has been classified from a physical and biological point of view as Physical classification of soil water, and biological classification of soil water.

Physical Classification of Soil Water

1. Gravitational water: Gravitational water occupies the larger soil pores (macro pores) and moves down readily under the force

of gravity. Water in excess of the field capacity is termed gravitational water. Gravitational water is of no use to plants because it occupies the larger pores. It reduces aeration in the soil. Thus, its removal from soil is a requisite for optimum plant growth. Soil moisture tension at gravitational state is zero or less than 1/3 atmosphere.

Factors Affecting Gravitational Water

i. Texture: Plays a great role in controlling the rate of movement of gravitational water. The flow of water is proportional to the size of particles. The bigger the particle, the more rapid is the flow or movement. Because of the larger size of pore, water percolates more easily and rapidly in sandy soils than in clay soils.

ii. Structure: It also affects gravitational water. In platy structure movement of gravitational water is slow and water stagnates in the soil. Granular and crumby structure helps to improve gravitational water movement. In clay soils having single grain structure, the gravitational water, percolates more slowly. If clay soils form aggregates (granular structure), the movement of gravitational water improves.

2. Capillary water: Capillary water is held in the capillary pores (micro pores). Capillary water is retained on the soil particles by surface forces. It is held so strongly that gravity cannot remove it from the soil particles. The molecules of capillary water are free and mobile and are present in a liquid state. Due to this reason, it evaporates easily at ordinary temperature though it is held firmly by the soil particle; plant roots are able to absorb it. Capillary water is, therefore, known as available water. The capillary water is held between 1/3 and 31 atmosphere pressure.

Factors Affecting Capillary Water: The amount of capillary water that a soil is able to hold varies considerably. The following factors are responsible for variation in the amount of capillary water.

i. Surface tension: An increase in surface tension increases the amount of capillary water.

ii. Soil texture: The finer the texture of a soil, greater is the amount of capillary water holds. This is mainly due to the greater surface area and a greater number of micro pores.

iii. Soil structure: Platy structure contains more water than granular structure.

iv. Organic matter: The presence of organic matter helps to increase the capillary capacity of a soil. Organic matter itself has a great capillary capacity. Undecomposed organic matter is generally porous having a large surface area, which helps to hold more capillary water. The humus that is formed on decomposition has a great capacity for absorbing and holding water. Hence the presence of organic matter in soil increases the amount of capillary water in soil.

3. Hygroscopic water: The water that held tightly on the surface of soil colloidal particle is known as hygroscopic water. It is essentially non-liquid and moves primarily in the vapour form.

Hygroscopic water held so tenaciously (31 to 10000 atmospheres) by soil particles that plants can not absorb it. Some microorganism may utilize hygroscopic water. As hygroscopic water is held tenaciously by surface forces its removal from the soil requires a certain amount of energy. Unlike capillary water which evaporates easily at atmospheric temperature, hygroscopic water cannot be separated from the soil unless it is heated.

Factors Affecting Hygroscopic Water: Hygroscopic water is held on the surface of colloidal particles by the dipole orientation of water molecules. The amount of hygroscopic water varies inversely with the size of soil particles. The smaller the particle, the greater is the amount of hygroscopic water it adsorbs. Fine textured soils like clay contain more hygroscopic water than coarse textured soils.

The amount of clay and also its nature influences the amount of hygroscopic water. Clay minerals of the montmorillonite type with their large surface area adsorb more water than those of the kaolinite type, while illite minerals are intermediate.

Biological Classification of Soil Water

There is a definite relationship between moisture retention and its utilization by plants. This classification based on the availability of water to the plant. Soil moisture can be divided into three parts.

i. Available water: The water which lies between wilting coefficient and field capacity. It is obtained by subtracting wilting coefficient from moisture equivalent.

ii. Unavailable water: This includes the whole of the hygroscopic water plus a part of the capillary water below the wilting point.

iii. Super available or superfluous water: The water beyond the field capacity stage is said to be super available. It includes gravitational water plus a part of the capillary water removed from larger interstices. This water is unavailable for the use of plants. The presence of super-available water in a soil for any extended period is harmful to plant growth because of the lack of air.

Soil Moisture Constants

Earlier classification divided soil water into gravitational, capillary and hygroscopic water. The hygroscopic and capillary waters are in equilibrium with the soil under given condition. The hygroscopic coefficient and the maximum capillary capacity are the two equilibrium points when the soil contains the maximum amount of hygroscopic and capillary waters, respectively. The amount of water that a soil contains at each of these equilibrium points is known as soil moisture constant.

The soil moisture constant, therefore, represents definite soil moisture relationship and retention of soil moisture in the field. The three classes of water (gravitational, capillary and hygroscopic) are however very broad and do not represent accurately the soil - water relationships that exists under field conditions.

1. Though the maximum capillary capacity represents the maximum amount of capillary water that a soil holds, the whole of capillary water is not available for the use of the plants.
2. A part of it, at its lower limit approaching the hygroscopic coefficient is not utilized by the plants.
3. Similarly a part of the capillary water at its upper limit is also not available for the use of plants.

Hence two more soil constants, viz., field capacity and wilting coefficient have been introduced to express the soil-plant-water relationships as it is found to exist under field conditions.

1. Field capacity: Assume that water is applied to the surface of a soil. With the downward movement of water all macro and micro pores are filled up. The soil is said to be saturated with respect to water and is at maximum water holding capacity or maximum retentive capacity. It is the amount of water held in the soil when all pores are filled.

   Sometimes, after application of water in the soil all the gravitational water is drained away, and then the wet soil is almost uniformly moist. The amount of water held by the soil at this stage is known as the field capacity or normal moisture capacity of that soil. It is the capacity of the soil to retain water against the downward pull of the force of gravity. At this stage only micropores or capillary pores are filled with water and plants absorb water for their use. At field capacity water is held with a force of 1/3 atmosphere. Water at field capacity is readily available to plants and microorganism

2. Wilting coefficient: As the moisture content falls, a point is reached when the water is so firmly held by the soil particles that plant roots are unable to draw it. The plant begins to wilt. At this stage even if the plant is kept in a saturated atmosphere it does not regain its turgidity and wilts unless water is applied to the soil. The stage at which this occurs is termed the Wilting point and the percentage amount of water held by the soil at this stage is known as the Wilting Coefficient. It represents the point at which the soil is unable to supply water to the plant. Water at wilting coefficient is held with a force of 15 atmospheres.
3. Hygroscopic coefficient: The hygroscopic coefficient is the maximum amount of hygroscopic water absorbed by 100 g of dry soil under standard conditions of humidity (50% relative humidity) and temperature (15°C). This tension is equal to a force of 31 atmospheres. Water at this tension is not available to plant but may be available to certain bacteria.
4. Available water capacity: The amount of water required to apply to a soil at the wilting point to reach the field capacity is called the "available" water. The water supplying power of soils is related to the amount of available water a soil can hold. The available water is the difference in the amount of water at field capacity (- 0.3 bar) and the amount of water at the permanent wilting point (- 15 bars).
5. Maximum water holding capacity: It is also known as maximum retentive capacity. It is the amount of moisture in a soil when its pore spaces both micro and macro capillary are completely filled with water. It is a rough measure of total pore space of soil. Soil moisture tension is very low between 1/100th to 1/1000th of an atmosphere or pF 1 to 0.
6. Sticky point moisture: It represents the moisture content of soil at which it no longer sticks to a foreign object. The sticky

point represents the maximum moisture content at which a soil remains friable. Sticky point moisture values vary nearly approximate to the moisture equivalent of soils. Summary of the soil moisture constants, type of water and force with which it held is given in following table.

Soil moisture constants and range of tension and PF

Sr. No.	*Moisture class*	*Range of tension in atmosphere*	*Equivalent PF range*
1	Chemically combined	Very High	-
2	Water vapour	Held at saturation point in the soil air	-
3	Hygroscopic	31 to 10,000	4.50 to 7.00
4	Hygroscopic coefficient	31	4.50
5	Wilting point	15	4.20
6	Capillary	1/3 to 31	2.54 to 4.50
7	Moisture equivalent	1/3 to 1	2.70 to 3.00
8	Field capacity	1/3	2.54

Entry of Water Into Soil

1. Infiltration: Infiltration refers to the downward entry or movement of water into the soil surface. It is a surface characteristic and hence primarily influenced by the condition of the surface soil. Soil surface with vegetative cover has more infiltration rate than bare soil. Warm soils absorb more water than colder ones. Coarse surface texture, granular structure and high organic matter content in surface soil, all help to increase infiltration. Infiltration rate is comparatively lower in wet soils than dry soils.

Factors affecting Infiltration

1. Clay minerals
2. Soil Texture
3. Soil structure
4. Moisture content
5. Vegetative cover
6. Topography

2. Percolation: The movement of water through a column of soil is called percolation. It is important for two reasons.

i) This is the only source of recharge of ground water which can be used through wells for irrigation

ii) Percolating waters carry plant nutrients down and often out of reach of plant roots (leaching)

Percolation is dependent of rainfall. In dry region it is negligible and under high rainfall it is high. Sandy soils have greater percolation than clayey soil. Vegetation and high water table reduce the percolation loss

3. Permeability: It indicates the relative ease of movement of water with in the soil. The characteristics that determine how fast air and water move through the soil are known as permeability. The term hydraulic conductivity is also used which refers to the readiness with which a soil transmits fluids through it.

Soil Water Movement

i) Saturated Flow

ii) Unsaturated Flow

iii) Water Vapour Movement

Saturated Flow: This occurs when the soil pores are completely filled with water. This water moves at water potentials larger than – 33 k Pa. Saturated flow is water flow caused by gravity's pull. It begins with *infiltration,* which is water movement into soil when rain or irrigation water is on the soil surface. When the soil profile is wetted, the movement of more water flowing through the wetted soil is termed *percolation.*

Hydraulic conductivity can be expressed mathematically as

$V = kf$

Where,

V = Total volume of water moved per unit time

f = Water moving force

k = Hydraulic conductivity of soil

Factors affecting movement of water

1. Texture, 2.Structure, 3.Amount of organic matter, 4.Depth of soil to hard pan, 5.Amount of water in the soil, 6.temperature and 7. Pressure

Vertical Water Flow: The vertical water flow rate through soil is given by *Darcy's law*. The law states that the rate of flow of liquid or flux through a porous medium is proportional to the hydraulic gradient in the direction of floe of the liquid.

$$QW = - k \frac{(dw)\ At}{Ds}$$

Where,

QW = Quantity of water in cm-3

k = rate constant (cm/s)

dw = Water height (head), cm

A = Soil area (cm2))

t = Time

ds = Soil depth (cm)

Soil Air

Soil air is a continuation of the atmospheric air. Unlike the other components, it is constant state of motion from the soil pores into the atmosphere and from the atmosphere into the pore space. This constant movement or circulation of air in the soil mass resulting in the renewal of its component gases is known as soil aeration.

Composition of Soil Air: The soil air contains a number of gases of which nitrogen, oxygen, carbon dioxide and water vapour are the most important. Soil air constantly moves from the soil pores into the atmosphere and from the atmosphere into the pore space. Soil air and atmospheric air differ in the compositions. Soil air contains a much greater proportion of carbon dioxide and a lesser amount of oxygen than atmospheric air. At the same time, soil air contains a far great amount of water vapour than atmospheric air. The amount of nitrogen in soil air is almost the same as in the atmosphere.

Composition of Soil and Atmospheric Air

	Percentage by volume		
	Nitrogen	***Oxygen***	***Carbon dioxide***
Soil Air	79.2	20.6	0.3
Atmospheric Air	79.9	20.97	0.03

Factors Affecting the Composition of Soil Air:

1. Nature and condition of soil: The quantity of oxygen in soil air is less than that in atmospheric air. The amount of oxygen also depends upon the soil depth. The oxygen content of the air in lower layer is usually less than that of the surface soil. This is possibly due to more readily diffusion of the oxygen from the atmosphere into the surface soil than in the subsoil. Light texture soil or sandy soil contains much higher percentage than heavy soil. The concentration of CO2 is usually greater in subsoil probably due to more sluggish aeration in lower layer than in the surface soil.

2. Type of crop: Plant roots require oxygen, which they take from the soil air and deplete the concentration of oxygen in the soil air. Soils on which crops are grown contain more CO_2 than fallow lands. The amount of CO_2 is usually much greater near the roots of plants than further away. It may be due to respiration by roots.
3. Microbial activity: The microorganisms in soil require oxygen for respiration and they take it from the soil air and thus deplete its concentration in the soil air. Decomposition of organic matter produces CO_2 because of increased microbial activity. Hence, soils rich in organic matter contain higher percentage of CO_2.
4. Seasonal variation: The quantity of oxygen is usually higher in dry season than during the monsoon. Because soils are normally drier during the summer months, opportunity for gaseous exchange is greater during this period. This results in relatively high O_2 and low CO_2 levels. Temperature also influences the CO_2 content in the soil air. High temperature during summer season encourages microorganism activity which results in higher production of CO_2.

Exchange of Gases between Soil and Atmosphere

The exchange of gases between the soil and the atmosphere is facilitated by two mechanisms

1. Mass flow: With every rain or irrigation, a part of the soil air moves out into the atmosphere as it is displaced by the incoming water. As and when moisture is lost by evaporation and transpiration, the atmospheric air enters the soil pores. The variations in soil temperature cause changes in the temperature of soil air. As the soil air gets heated during the day, it expands and the expanded air moves out into the atmosphere. On the other hand, when the soil begins to cool, the soil air contracts and the atmospheric air is drawn in.
2. Diffusion: Most of the gaseous interchange in soils occurs by diffusion. Atmospheric and soil air contains a number of gases such as nitrogen, oxygen, carbon dioxide etc., each of which exerts its own partial pressure in proportion to its concentration.

The movement of each gas is regulated by the partial pressure under which it exists. If the partial pressure on one of the gases (i.e. carbon dioxide) is greater in the soil air than in the atmospheric air, it (CO_2) moves out into the atmosphere. Hence, the concentration of CO_2 is more in soil air.

On the other hand, partial pressure of oxygen is low in the soil air, as oxygen present in soil air is consumed as a result of biological activities. The oxygen present in the atmospheric air (partial pressure of O2 is greater) therefore, diffuses into the soil air till equilibrium is established. Thus, diffusion allows extensive movement and continual change of gases between the soil air and the atmospheric air. Oxygen and carbon dioxide are the two important gases that take in diffusion

Importance of Soil Aeration

1. Plant and root growth: Soil aeration is an important factor in the normal growth of plants. The supply of oxygen to roots in adequate quantities and the removal of CO2 from the soil atmosphere are very essential for healthy plant growth.

When the supply of oxygen is inadequate, the plant growth either retards or ceases completely as the accumulated CO2 hampers the growth of plant roots. The abnormal effect of insufficient aeration on root development is most noticeable on the root crops. Abnormally shaped roots of these plants are common on the compact and poorly aerated soils. The penetration and development of root are poor. Such undeveloped root system cannot absorb sufficient moisture and nutrients from the soil

2. Microorganism population and activity: The microorganisms living in the soil also require oxygen for respiration and metabolism. Some of the important microbial activities such as the decomposition of organic matter, nitrification, Sulphur oxidation etc depend upon oxygen present in the soil air. The deficiency of air (oxygen) in soil slows down the rate of microbial activity.

For example, the decomposition of organic matter is retarded and nitrification arrested. The microorganism population is also drastically affected by poor aeration.

3. Formation of toxic material: Poor aeration results in the development of toxin and other injurious substances such as ferrous oxide, H2S gas, CO2 gas etc in the soil.
4. Water and nutrient absorption: A deficiency of oxygen has been found to check the nutrient and water absorption by plants. The energy of respiration is utilized in absorption of water and nutrients. Under poor aeration condition (this condition may arise when soil is water logged), plants exhibit water and nutrient deficiency

5. Development of plant diseases: Insufficient aeration of the soil also leads to the development of diseases. For example, wilt of gram and dieback of citrus and peach.

Organic Colloids – Soil Organic Matter

Soil organic matter (SOM) comprises an accumulation of

i) Partially disintegrated and decomposed plant and animal residues

ii) Other organic compounds synthesized by the soil microbes upon decay.

OM content of a well drained mineral soil is LOW: 1 – 6 % by weight in the top soil and even less in the subsoil.

Sources of Soil Organic Matter

The primary sources of SOM are plant tissues

1. The tops and roots of trees
2. Shrubs, grasses, remains of harvested crops and Soil organisms

Animals are secondary sources of Organic Matter.

1. Waste products of animals
2. Remains of animals after completion of life cycle.

Factors Affecting Soil Organic Matter

1. Climate: Temperature and rainfall exert a dominant influence on the amounts of N and OM found in soils.
   a) Temperature: The OM and N content of comparable soils tend to increase if one moves from warmer to cooler areas. The decomposition of OM is accelerated in warm climates as compared to cooler climates. For each 10 C decline in mean annual temperature, the total OM and N increases by two to three times.
   b) Rainfall: There is an increase in OM with an increase in rainfall. Under comparable conditions, the N and OM increase as the effective moisture becomes greater.
2. Natural Vegetation: The total OM is higher in soils developed under grasslands than those under forests.
3. Texture: Fine textured soils are generally higher in OM than coarse textured soils.
4. Drainage: Poorly drained soils because of their high moisture content and relatively poor aeration are much higher in OM and N than well drained soils.

5. Cropping and Tillage: The cropped lands have much low N and OM than comparable virgin soils. Modern conservation tillage practices helps to maintain high OM levels as compared to conventional tillage.
6. Rotations, residues and plant nutrients: Crop rotations of cereals with legumes results in higher soil OM. Higher OM levels, preferably where a crop rotation is followed.

Decomposition of Soil Organic Matter

The organic materials (plant and animal residues) incorporated in the soil are attacked by a variety of microbes, worms and insects in the soil if the soil is moist. Some of the constituents are decomposed very rapidly, some less readily, and others very slowly. The list of constituents in terms of ease of decomposition:

1. Sugars, starches and simple proteins Rapid Decomposition
2. Crude proteins
3. Hemicelluloses
4. Cellulose
5. Fats, waxes, resins
6. Lignins Very slow Decomposition

The organic matter is also classified on the basis of their rate of decomposition

Rapidly decomposed: Sugars, starches, proteins etc

Less rapidly decomposed: Hemicelluloses, celluloses etc

Very slowly decomposed: Fats, waxes, resins, lignins etc

Simple decomposition products

Aerobic – CO2, H2O, NO3, SO4

When organic material is added to soil, three general reactions take place

a. The bulk of the material undergoes enzymatic oxidation with CO2, water, energy and heat as the major products.
b. The essential elements such as N, P and S are released and / or immobilized by a series of reactions.

Compounds very resistant to microbial action is formed either through modification of compounds or by microbial synthesis.

A. Decomposition of soluble substances: When glucose is decomposed under aerobic conditions the reaction is as under:

Sugar + Oxygen ——————> $CO_2 + H_2O$

Under partially oxidized conditions,

Sugar + Oxygen ——————> Aliphatic acids
(Acetic, formic etc.) or
Hydroxy acids
(Citric, lactic etc.) or
Alcohols (ethyl alcohol etc.)

Some of the reactions invoiced may be represented as under:

$C_6H_{12}O_6 + 2O_2$ ——————> $2\ CH_3.COOH + 2CO_2 + 2H_2O$

$2C_6H_{12}O_6 + 3O_2$ ——————> $2\ C_6H_8O_7 + 4\ H_2O$

$C_6H_{12}O_6 + 2O_2$ ——————> $2C_2H_5OH + 2\ CO_2$

i) Ammonification: The transformation of organic nitrogenous compounds (amino acids, amides, ammonium compounds, nitrates etc.) into ammonia is called ammonification. This process occurs as a result of hydrolytic and oxidative enzymatic reaction under aerobic conditions by heterotrophic microbes.

ii) Nitrification: The process of conversion of ammonia to nitrites (NO_2) and then to nitrate (NO_3^-) is known as nitrification. It is an aerobic process by autotrophic bacteria.

Nitrosomonas Nitrobacter

NH_3 ——————> NO_2 ——————> NO_3^-

Ammonia Nitrite Nitrate

The net reactions are as follows:

$NH_4 + O_2$ ——————> $NO_2 + 2H^+ + H_2O$ + energy

$NO_2 + O_2$ —————— >NO_3^- + energy

iii) Denitrification: The process, which involves conversion of soil nitrate into gaseous nitrogen or nitrous oxide, is called Denitrification. Water logging and high pH will increase N loss by Denitrification.

Pseudomonas / Bacillus

Nitrate ——————————> Nitrogen Gas

Decomposition of Insoluble Substances

i) Breakdown of Protein: During the course of decomposition of plant materials, the proteins are first hydrolyzed to a number of intermediate products and may be represented as under:

Hydrolysis Proteases Aas

Proteins ———> Peptones ———> Amides ——> Ammonia Peptides

Aminization Ammonification

Aminization: The process of conversion of proteins to aminoacids.

Ammonification: The process of conversion of aminoacids and amides to ammonia.

ii) Breakdown of cellulose: The decomposition of the most abundant carbohydrates is as follows:

hydrolysis hydrolysis

Cellulose ——————> Cellobiose ——————> Glucose

(cellulase) (cellobiase)

oxidation

——————> Organic acids ——————> $CO_2 + H_2O$

This reaction proceeds more slowly in acid soils than in neutral and alkaline soils. It is quite rapid in well aerated soils and comparatively slow in poorly aerated soils.

iii) Breakdown of Hemicellulose: Decompose faster than cellulose and are first hydrolyzed to their components sugars and uronic acids. Sugars are attacked by microbes and are converted to organic acids, alcohols, carbon dioxide and water. The uronic acids are broken down to pentose and CO_2. The newly synthesized hemicelluloses thus form a part of the humus.

iv) Breakdown of Starch: It is chemically a glucose polymer and is first hydrolyzed to maltose by the action of amylases. Maltose is next converted to glucose by maltase. The process is represented as under:

$(C_6H_{10}O_5)_n + nH_2O$ ——————> $(C_6H_{12}O_6)$

C. Decomposition of ether soluble substances:

Fats————————> glycerol + fatty acids

Glycerol ——————> CO_2 + water

Decomposition of Lignin: Lignin decomposes slowly, much slower than cellulose. Complete oxidation gives rise to CO_2 and H_2O.

Factors affecting decomposition of organic matter

1. Temperature: Cold periods retard plant growth and OM decomposition. Warm summers may permit plant growth and humus accumulation.

2. Soil moisture: Extremes of both arid and anaerobic conditions reduce plant growth and microbial decomposition. Near or slightly wetter than field capacity moisture conditions are most favourable for both processes.
3. Nutrients: Lack of nutrients particularly N slows decomposition.
4. Soil PH: Most of the microbes grow best at pH 6 – 8, but are severely inhibited below pH 4.5 and above pH 8.5.
5. Soil Texture: Soils higher in clays tend to retain larger amounts of humus.
6. Other Factors: Toxic levels of elements (Al, Mn, B, Se, Cl), excessive soluble salts, shade and organic phytotoxins in plant materials.

Role of Organic Matter

1. Organic Matter creates a granular condition of soil which maintains favourable condition of aeration and permeability.
2. Water holding capacity of soil is increased and surface runoff, erosion etc., are reduced as there is good infiltration due to the addition of OM.
3. Surface mulching with coarse OM lowers wind erosion and lowers soil temperatures in the summer and keeps the soil warmer in winter.
4. OM serves as a source of energy for the microbes and as a reservoir of nutrients that are essential for plant growth and also hormones, antibiotics.
5. Fresh OM supplies food for earthworms, ants and rodents and makes soil P readily available in acid soils.
6. Organic acids released from decomposing OM help to reduce alkalinity in soils; organic acids along with released CO2 dissolve minerals and make them more available.
7. Humus (a highly decomposed OM) provides a storehouse for the exchangeable and available cations.
8. It acts as a buffering agent which checks rapid chemical changes in pH and soil reaction.

Humus

Humus is a complex and rather resistant mixture of brown or dark brown amorphous and colloidal organic substance that results from microbial decomposition and synthesis and has chemical and physical properties of great significance to soils and plants.

Humus Formation

The humus compounds have resulted from two general types of biochemical reactions: Decomposition and Synthesis.

1. Decomposition:
   a) Chemicals in the plant residues are broken down by soil microbes including lignin.
   b) Other simpler organic compounds that result from the breakdown take part immediately in the second of the humus-forming processes, biochemical synthesis.
   c) These simpler chemicals are metabolized into new compounds in the body tissue of soil microbes.
   d) The new compounds are subject to further modification and synthesis as the microbial tissue is subsequently attacked by other soil microbes.
2. Synthesis: Involve such breakdown products of lignin as the phenols and quinones.
   a) These monomers undergo polymerization by which polyphenols and polyquinones are formed.
   b) These high molecular weight compounds interact with N-containing amino compounds and forms a significant component of resistant humus.
   c) Colloidal clays encourage formation of these polymers.
   d) Generally two groups of compounds that collectively make up humus, the humic group and the nonhumic group.

Properties of Humus

1. The tiny colloidal particles are composed of C, H, and O2. The colloidal particles are negatively charged (-OH, -COOH or phenolic groups), has very high surface area, higher CEC (150 – 300 cmol/kg), 4 - 5 times higher WHC than that of silicate clays.
2. Humus has a very favourable effect on aggregate formation and stability.
3. Impart black colour to soils.
4. Cation exchange reactions are similar to those occurring with silicate clays.

Clay – Humus Complex

Humus, the organic amorphous colloid supplies both basic and acidic ions which is transitory and ultimately disappears from soil.

Clay, the inorganic crystalline colloid supplies chiefly the basic nutrient ions is more or less stable. Both these colloids form the soil colloidal complex and are extremely active and form important sources of plant nutrients.

It is believed that humus and clay exist in the soil as clay – humus complex, the two being held together by cations like Ca, Fe, etc. Depending upon the nature of binding cation, two types of Clay – humus complex have been recognized. The colloidal complex bound by Ca ions is more stable and is responsible for the favourable physical condition of the soil, particularly its structure. The other type where Fe acts as the binding agent creates a poor physical condition of the soils.

Ion Exchange

As soils are formed during the weathering processes, some minerals and organic matter are broken down to extremely small particles. Chemical changes further reduce these particles until they cannot be seen with the naked eye. The very smallest particles are called colloids.

The mineral clay colloids are plate like in structure and crystalline in nature. In most soils; clay colloids exceed organic colloids in amount.

Colloids are primarily responsible for the chemical reactivity in soils. The kind of parent material and the degree of weathering determine the kinds of clays present in the soil. Since soil colloids are derived from these clays, their reactivity is also influenced by parent material and weathering. Each colloid (inorganic and organic) has a net negative (-) charge developed during the formation process. This means it can attract and hold positively (+) charged particles. An element with an electrical charge is called an ion.

Potassium, sodium (Na), hydrogen (H), Ca and Mg all has positive charges. They are called cations and ions with negative charges, such as nitrate and sulphate, are called anions.

Negatively charged colloids attract cations and hold them like a magnet holds small pieces of metal. This characteristic explains; why nitrate-N is more easily leached from the soil than ammonium-N. Nitrate has a negative charge like soil colloids. So, NO3- is not held by the soils, but remains as a free ion in soil water to be leached through the soil profile in some soils and under some rainfall conditions. The charges associated with soil particles attract simple and complex ions of opposite charge. Thus, a given colloidal mixture may exhibit not only a maze of positive and negative surface charges but also an equal complex complement of simple cations and anions such as Ca2+ and S04 - that are attracted by the particle charges.

The adsorbed anions are commonly present in smaller quantities than the cations because the negative charges generally predominate on the soil colloid.

Mechanism of Cation Exchange: The exchange of cations has been explained on the basis of the electro-kinetic theory of ion exchange. According to this theory, the adsorbed cations forming the outer shell of the ionic double layer are supposed to be in a state of oscillation when suspended in water, forming a diffuse double layer. Due to these oscillations, some of the cations move away from the surface of the clay micelle. In the presence of the solution of an electrolyte a cation of the added electrolyte slips in between the inner negative layer and the outer oscillating positive ion. The electrolyte cation is now adsorbed on the micelle and the surface cation remains in solution as an exchanged ion. Thus the exchange of cations takes place. Cations are positively charged nutrient ions and molecules. While, clay particles are negatively charged constituents of soils. These negatively charged particles (clay) attract, hold and release positively charged nutrient ions (cations). Organic matter particles also have a negative charge to attract cations. Sand particles carry little or no charge and do not react. Cations held by soils can be replaced by other cations. This means they are exchangeable. For example, Ca++ can be exchanged for H+ and /or K+ and vice versa.

Cation Exchange Capacity (CEC): The CEC is the capacity of soil to hold and exchange cations. The cation exchange capacity is defined simply as the sum total of the exchangeable cations that a soil can adsorb. The higher the CEC of soil the more cations it can retain. Soils differ in their capacities to hold exchangeable K+ and other cations.

The CEC depends on amount and kinds of clay and organic matter present. A high-clay soil can hold more exchangeable cations than a low-clay soil. CEC also increases as organic matter increases. Clay minerals usually range from 10 to 150 meq/100 g in CEC values. Organic matter ranges from 200 to 400 meq/100 g. So, the kind and amount of clay and organic matter content greatly influence the CEC of soils. Clay soils with high CEC can retain large amounts of cations against potential loss by leaching. Sandy soils, with low CEC, retain smaller quantities.

This makes timing and application rates important in planning a fertilizer programme. For example, it may not be wise to apply K on very sandy soils in the middle of a monsoon, where rainfall can be high and intense. Fertilizer application should be split to prevent leaching and losses through erosion. Also, splitting N applications to meet peak crop demand are important to reduce the potential for nitrate leaching on sands as well as finer-textured soils.

Means of Expression: The cation exchange capacity is expressed in terms of equivalents or more specifically, as milli equivalents per 100 gram and is written as meq /100g. The term equivalent is defined as one gram of atomic weight of hydrogen (or the amount of any other ion) that will combine with or displace this amount of hydrogen for monovalent ions such as Na+, K+, NH4+ and Cl -, the equivalent weight and atomic weight are same, since they can replace one H ion. Divalent cations such as Ca++ and Mg++ can take the place of two H+ ions. The milliequivalent weight of a substance is one thousandth of its atomic weight. Since the atomic weight of hydrogen is about 1 gram. The term milliequivalent (meq.) may be defined as 1 milligram of hydrogen.

This unit of exchangeable cations i.e. milliequivalent per 100 g of soil (meq/100g) was used prior to 1982. In the newer metric system the term equivalent is not used, however, now moles are the accepted chemical unit. All the calculation and concepts of "equivalents" are still mentally used but the notation must be written differently. The old "equivalent" is represented by moles (+) or mole, which indicates a monovalent ion portion. For example, to write 12.5 meq/100 g in the newer metric system, it can be written as: 12.5 c mol (+) kg-1 of soil (centimoles) or 125 m mol (+) kg-1 of soil (millimoles).

Replacing Power of Cations: The replacing power of cations varies with the type of ion, its size and degree of hydration, valence and concentration and the kind of clay mineral involved, as it is controlled by number of factors no single order of replacement can be given. All other factors being equal the replacing power of monovalent cations increases in the following order: Li < Na < K < Rb < Cs < H and for divalent cations: Mg < Ca < Sr < Ba. In case of mixture of monovalent and divalent cations as they exist in normal soils the replacing power increases in the following order: Na < K < NH4 < Mg < Ca < H. This means Na is more easily replaced than K and K more easily than NH4 and so on.

Percent Base Saturation: The percent of total CEC occupied by the major cations has been used in the past to develop fertilizer programmes. The idea is that certain nutrient ratios or 'balances' are needed to ensure proper uptake by the crop for optimum yields. Research has shown, however, that cation saturation ranges and ratios have little or no utility in a vast majority of agricultural soils. Under field conditions, ranges of nutrients can vary widely with no detrimental effects, so long as individual nutrients are present in sufficient levels in the soil to support optimum plant growth.

Importance of Cation Exchange: Cation exchange is an important reaction in soil fertility, in causing and correcting soil acidity and basicity, in changes altering soil physical properties, and as a mechanism in purifying or altering percolating waters. The plant nutrients like calcium, magnesium, and potassium are supplied to plants in large measure from exchangeable forms.

Soil Reaction

Soil reaction is one of the most important physiological characteristics of the soil solution. The presence and development of micro- organisms and higher plants depend upon the chemical environment of soil. There fore study of soil reaction is important in soil science.

There are three Types of Soil Reactions: 1. Acidic 2. Alkaline and 3. Neutral

1. Acidic: It is common in region where precipitation is high. The high precipitation leaches appreciable amounts of exchangeable bases from the surface layers of the soils so that the exchange complex is dominated by H ions. Acid soils, therefore, occur widely in humid regions and affect the growth of plants markedly.
2. Alkaline: Alkali soils occur when there is comparatively high degree of base saturation. Salts like carbonates of calcium, magnesium and sodium also give a preponderance of OH ions over H ions in the soil solution. When salts of strong base such as sodium carbonate go into soil solution and hydrolyze, consequently they give rise to alkalinity. The reaction is as follows:

Na2CO3 ——à 2Na + + CO3=

2Na+ + CO3= + 2HOH ——à 2Na+ + 2OH - + H2CO3

Since sodium hydroxide dissociates to a greater degree than the carbonic acid, OH ions dominate and give rise to alkalinity. This may be as high as 9 or 10. These soils most commonly occur in arid and semi-arid regions.

3. Neutral: Neutral soils occur in regions where H ions just balance OH ions.

Soil pH: The reaction of a solution represents the degree of acidity or basicity caused by the relative concentration of H ions (acidity) or OH ions present in it. Acidity is due to the excess of H ions over OH ions, and alkalinity is due to the excess of OH ions over H ions. A neutral reaction is produced by an equal activity of H and OH ions. According to the theory of dissociation, the activity is due to the dissociation or ionization of compounds into ions.

Method of Expressing Acidity or Alkalinity

Equivalent quantities of all acids or alkalies contain the same number of total H or OH ions, respectively. But, when they are dissolved in water they do not ionize to the same extent. The amount of acid or alkali ionized depends upon the content of free H and OH ions. When the dissociation is high e.g., hydrochloric acid (a strong acid), it dissociates to a larger extent than the weak acetic acid. Acetic acid dissociates only to about 10 % as compared to hydrochloric acid. In a 1N solution of hydrochloric acid there will be 1 gram of H ions per litre, while in a normal solution of acetic acid there will be 1/10 gram of H ions per litre. But in titration 1 ml of 1 N hydrochloric acid and 1 ml of 1 N acetic acid will require 1 ml of 1 N sodium hydroxide for neutralization separately, because the total acidity is the same and titration determines both the ionized and unionized H or OH ions.

Soil Acidity: There are three kinds of acidity.

(i) Active acidity is due to the H^+ ion in the soil solution.

(ii) Salt replaceable acidity represented by the hydrogen and aluminum that are easily exchangeable by other cations in a simple unbuffered salt solution such as KCl and

(iii) Residual acidity is that acidity which can be neutralized by limestone or other alkaline materials but cannot be detected by the salt-replaceable technique. Obviously, these types of acidity all add up to the total acidity of a soil.

i. Active acidity: The active acidity is a measure of the H^+ ion activity in the soil solution at any given time. However, the quantity of H^+ ions owing to active acidity is very small compared to the quantity in the exchange and residual acidity forms. For example, only about 2 kg of calcium carbonate would be required to neutralize the active acidity in a hectare-furrow slice of an average mineral soil at pH

4 and 200/0 moisture. Even though the concentration of hydrogen ions owing to active acidity is extremely small, it is important because this is the environment to which plants and microbes are exposed.

ii. Salt replaceable (exchangeable) acidity: This type of acidity is primarily associated with the exchangeable aluminum and hydrogen ions that are present in largest quantities in very acid soils. These ions can be released into the soil solution by an unbuffered salt such as KCl.

Al3+ + 4KCI +AlCI3+HCI' L~~~ H + L~~~~ ~ (Soil Solid) (Soil Solution) (Soil Solid) (Soil solution), in moderately acid soils, the quantity of easily exchangeable aluminum and hydrogen is quite limited. Even in these soils, however, the limestone needed to neutralize this type of acidity is commonly more than 100 times that needed for the soil solution (active acidity). At a given pH value, exchangeable acidity is generally highest for smectites, intermediate for vermiculites, and lowest for kaolinite. In any case, however, it accounts for only a small portion of the total soil acidity as the next section will verify.

iii. Residual acidity: Residual acidity is that which remains in the soil after active and exchange acidity has been neutralized. Residual acidity is generally associated with aluminum hydroxy ions and with hydrogen and aluminum atoms that are bound in non exchangeable forms by organic matter and silicate clays. If lime is added to a soil, the pH increases and the aluminum hydroxy ions are changed to uncharged gibbsite as follows. OH- OH- AI (OH) 2+ 7 AI (OH) 2+ 7 AI (OH) 3 In addition, as the pH increases bound hydrogen and aluminum can be released by calcium and magnesium In the lime materials [Ca (OH) 2 is used as an example of the reactive calcium liming material] The residual acidity is commonly far greater than either the active or salt replaceable acidity. Conservative estimates suggest that the residual acidity may be 1000 times greater than the soil solution or active acidity in a sandy soil and 50000 or even 100000 times greater in a clayey soil high in organic matter. The amount of ground limestone recommended to at least partly neutralize residual acidity is commonly 4-8 metric tons (Mg) per hectare furrow slice (1.8-3.6 tons/AFS).

It is obvious that the pH of the soil solution is only "the tip of the iceberg" in determining how much lime is needed. Buffering and

Soil Reaction Buffer action: Buffering refers to resistance to a change in pH. If 1 ml HCI (of 0.1 N) is added to one litre of pure distilled water of pH 7.0, the resulting solution would have a pH of about 5.0. If on the other hand, the same amount of acid is added to a litre of soil suspension the resulting change in pH would be very small. There is, a distinct resistance to change in pH. This power to resist a change in pH is called buffer action. A buffer solution is one which contains reserve acidity and alkalinity and does not change pH with small additions of acids or alkalies. Buffer capacity: The colloidal complex acts as a powerful buffer in the soil and does not allow rapid and sudden changes in soil reaction. Buffering depends upon the amount of colloidal material present in soil. Clay soils rich in organic matter are more highly buffered than sandy soils. Buffer capacity of the soil varies with its cation exchange capacity (C.E.C.). The greater the C.E.C. the greater will be its buffer capacity. Thus heavier the texture and the greater the organic matter content of a soil, the greater is the amount of acid or alkaline material required to change its pH. Importance of Buffering to Agriculture: Changes In soil reaction (pH) have a direct influence on the plants and it also affects the availability of plant nutrients. Deficiency of certain plant nutrients and excess availability of others in toxic amounts would seriously upset the nutritional balance in the soil. Buffering prevents sudden changes and fluctuation in soil pH, so it regulates the availability of nutrients and also checks direct toxic effect to plants.

Factors Controlling Soil Reactions

Soil reaction varies due to following factors

1. Nature of soil colloids: The colloidal particles of the soil influence soil reaction to a very greatest extent. When hydrogen (H+) ion forms the predominant adsorbed cations on clay colloids, the soil reaction becomes acid.
2. Soil solution: The soil solution carries a number of salts dissolved in capillary water. The cations of the salts intermingle with those of the diffuse double layer of the clay particle and increase the concentration. The concentration of cations in bulk of the solution is more or less (or nearly) the same as that near the particle surfaces. For an unsaturated soil (Clay), the more compact the layer the greater is the number of hydrogen ions dissociating into the solution. This increases the acidity of the soil solution or lowers its pH. Under field conditions, the concentration of salts varies with the moisture content of the

soil. The more dilute the solution, the higher the pH value. Hence the pH tends to drop as the soil gets progressively dry. Soil reaction is also influenced by the presence of CO2 in soil air. As the CO2 concentration increases, the soil pH falls and increases the availability of the nutrients. Under field conditions, plant roots and micro-organism liberate enough CO2 , which results in lowering the pH appreciably. This principle of increasing the concentration of CO2 in soil air is also used in the reclamation of alkali soils.

3. Climate: Rainfall plays important role in determining the reaction of soil. In general, soils formed in regions of high rainfall are acidic (low pH value), while those formed in regions of low rainfall are alkaline (high pH value).
4. Soil management: Cultural operations in general tend to increase soil acidity. They make an acid soil more acidic, and an alkaline soil less alkaline. As a result of constant cultivation, basic elements are lost from the soil through leaching and crop removal. This leads to change the soil reaction to the acid side.
5. Parent materials: Soils developed from parent material of basic rocks generally have higher pH than those formed from acid rocks (e.g. granite). The influence of parent material is not very important as it is completely masked by the climatic conditions under which the soil is developed.
6. Precipitation: As water from rainfall passes through the soil, basic nutrients such as calcium (Ca) and magnesium (Mg) are leached. They are replaced by acidic elements including Al, H and manganese (Mn). Therefore, soils formed under high rainfall conditions are more acid than those formed under arid conditions.
7. Decomposition of organic matter: Soil organic matter is continuously being decomposed by micro-organisms into organic acids, carbon dioxide (CO2) and water, forming carbonic acid. Carbonic acid, in turn, reacts with the Ca and Mg carbonates in the soil to form more soluble bicarbonates, which are leached away, leaving the soil more acid.
8. Native vegetation: Soils often become more acid when crops are harvested because of removal of bases. Type of crop determines the relative amounts of removal. For example, legumes generally contain higher levels of bases than do grasses. Calcium and Mg contents also vary according to the portion (s)

of the plant harvested. Many legumes release H ions into their rhizosphere when actively fixing atmospheric N2. The acidity generated can vary from 0.2 to 0.7 pH units per mole of fixed N.

9. Soil depth: Except in low rainfall areas, acidity generally increases with depth, so the loss of topsoil by erosion can lead to a more acid pH in the plough layer. The reason is that more subsoil is included in the plow layer as topsoil is lost. There are areas, however, where subsoil pH is higher than that of the topsoil.
10. Nitrogen fertilization: Nitrogen from fertilizer, organic matter, and manure and legume N fixation produces acidity. Nitrogen fertilization speeds up the rate at which acidity develops. At lower N rates, acidification rate is slow, but is accelerated as N fertilizer rates increase.
11. Flooding: The overall effect of submergence is an increase of pH in acid soils and a decrease in basic soils. Regardless of their original pH values, most soils reach pH of 6.5 to 7.2 within one month after flooding and remain at the level until dried. Consequently, liming is of little value in flooded rice production. Further, it can induce deficiencies of micronutrients such as zinc (Zn).

Bibliography

Akhil Baruah: *Advanced Morphology of Angiosperms*, Aavishkar, Delhi, 2008.

Arthur W. Gilbert, Mortier F. Barrus and Daniel Dean: *Growing and Breeding of Potatoes*, Asiatic Pub, Delhi, 2006.

Berger, P.L.: *Pyramids of Sacrifice: Political Ethics and Social Change*, New York, Basic Books, 1974.

Brooks D.: *Water: Local-Level Management*, Ottawa, International Development Research Centre, 2002.

Chadha K. L. and Pareek O. P.: *Advances in Horticulture: Fruit Crops,* New Delhi, Malhotra Publishing House, 1993.

Collymore L.: *Fruit Production in Barbados*, Port of Spain, Trinidad and Tobago, 1996.

Crucefix D.: *Avocado Variety Selection for Export Development,* Roseau, CARDI, 1996.

David White: *The Physiology and Biochemistry of Prokaryotes*, Oxford University Press, Delhi, 2007.

Diederichsen A.: *Coriander (Coriandrum sativum L.).* Rome, International Plant Genetic Resources Institute, 1996.

Ferentinos L.: *Proceeding of the Sustainable Taro Culture for the Pacific Conference*, Honolulu, HITAHR, 1993.

Fortuner, R.: *Nematode Identification and Expert System Technology,* New York, Plenum Press, 1988.

Godden G.: *Growing Citrus Trees*, Australia, Lothian Publishing Company Pvt. Ltd., 1988.

Hardy B.: *Biology and Agronomy of Forage Arachis*, Cali, International Centre for Tropical Agriculture, 1994.

Herminie Broedel Kitchen: *Soils and Crops : Diagnostic Techniques*, Satish Serial Publishing, Allahabad, 2004.

Huaman Z.: *Descriptors for Sweet Potato*, Rome, International Board for Plant Genetic Resources, 1991.

Jacquat Christiane: *Plants from the Markets of Thailand*, Bangkok, Duang Kamol, 1990.

Jeffers P.: *Evaluation of Four Onion Varieties in Montserrat*, Plymouth, CARDI, 1992.

Keshav Prasad Yadav: *Application of Morphometry in Geomorphology*, Radha Pub, Delhi, 2008.

Lorenzen, S.: *The Phylogenetic Systematics of Freeliving Nematodes*, London, The Ray Society, 1994.

Mitra S.: *Postharvest Physiology and Storage of Tropical and Subtropical Fruits*, Oxon, CABI, 1997.

Mondal M.S.: *Pollen Morphology and Systematic Relationship of the Family Polygonaceae*, BSI, Delhi, 1997.

Nobel, P. S.: *Physicochemical and Environmental Plant Physiology*, Academic Press, San Diego, 1999.

Oldham P.: *Cost of Production of Major Tree Crops in Dominica*, Roseau, Ministry of Agriculture, 1991.

Pemberton, R. W.: *Predictable Risk to Native Plants in Weed Biological Control,* Oecologia, 2000.

Pramila Tripathi: *Botanical Pesticides in the Management of Post-harvest Fruit Diseases*, Daya, Delhi, 2005.

Pramod P Mahulikar and Kshama M Chavan: *Botanicals as Ecofriendly Pesticides*, New India Pub Agency, Delhi, 2007.

Rao, S. Balaachandra: *Indian Mathematics and Astronomy*, Bangalore, Jnana Deep Publications, 1994.

Rutherford Lyn.: *A Gourmet's Book of Mushrooms & Truffles*, Sydney, Golden Press Pvt. Ltd., 1991.

Sanchez, P. A.: *Properties and Management of Soils in the Tropics*, John Wiley and Sons, New York, 1976.

Shivanand Tolanur: *Practical Soil Science and Agricultural Chemistry*, International Book Distributing, Delhi, 2004.

Thomas E.: *Fruit Production in St. Kitts and Nevis*, Port of Spain, IICA, 1996.

Urton, Gary: *The Social Life of Numbers*, Austin, University of Texas Press, 1997.

Vanderplank, J.E.: *Plant Diseases: Epidemics and Control,* New York, NY, USA, Academic Press, 1963.

White, G.F.: *Natural Hazards: Local, National, Global*, Oxford University Press, New York, 1974.

Yoshida T.: *Cultivation of Citrus Genetic Resources for Evaluation of Characteristics*, Tokyo, JICA, 1996.

Index

O

P

R

S

T

U

W

❑❑❑